MANAGING PURCHASING

MAKING THE SUPPLY TEAM WORK

THE NAPM PROFESSIONAL DEVELOPMENT SERIES

Michiel R. Leenders
Series Editor

Volume I
VALUE-DRIVEN PURCHASING
Managing the Key Steps in the Acquisition Process
Michiel R. Leenders
Anna E. Flynn

Volume II
MANAGING PURCHASING
Making the Supply Team Work
Kenneth H. Killen
John W. Kamauff

Volume III
VALUE-FOCUSED SUPPLY MANAGEMENT
Getting the Most Out of the Supply Function
Alan R. Raedels

Volume IV
PURCHASING FOR BOTTOM LINE IMPACT
Improving the Organization Through Strategic Procurement
Lisa M. Ellram
Laura M. Birou

MANAGING PURCHASING

MAKING THE SUPPLY TEAM WORK

Volume II
The NAPM Professional Development Series

Kenneth H. Killen
Cuyahoga Community College
and
John W. Kamauff
School of Business Administration
University of Western Ontario

Tempe, Arizona

IRWIN
Professional Publishing

Chicago • Bogotá • Boston • Buenos Aires • Caracas
London • Madrid • Mexico City • Sydney • Toronto

To those many purchasing professionals at the
Purchasing Management Association of Cleveland and the NAPM
—Thank you!
To those academies who have been and are my teachers,
mentors and colleagues
—Thank you!
To Robert L. Janson and Dr. Harold Puff
—a special thank you!

—Kenneth H. Killen

To Susan, Lexi, Tari, Ryan and Tess

—John W. Kamauff

Senior sponsoring editor: *Cynthia A. Zigmund*
Marketing manager: *J. D. Kinney*
Project editor: *Ethel Shiell/Montgomery Media, Inc.*
Production supervisor: *Pat Frederickson*
Art coordinator: *Montgomery Media, Inc.*
Compositor: *Montgomery Media, Inc.*
Typeface: *11/13 Times Roman*
Printer: *Buxton-Skinner*

Library of Congress Cataloging-in-Publication Data

Killen, Kenneth H.
Managing purchasing: making the supply team work/
Kenneth H. Killen, John W. Kamauff.
p. cm.—(The NAPM professional development series; bk. 2)
Based on: C.P.M. study guide. 6th ed. 1992
Includes index.
ISBN 0-7863-0127-9
1. Industrial procurement—Management. 2. Purchasing.
I. Kamauff, John W. II. C.P.M. study guide. III. Title.
IV. Series: NAPM professional development series; v. 2.
HD39.5.K53 1995
658.7 2—dc20 94-44896

Printed in the United States of America

1 2 3 4 5 6 7 8 9 0 BS 2 1 0 9 8 7 6 5

SERIES OVERVIEW

The fundamental premise for this series of four textbooks is that effective purchasing or supply management can contribute significantly to organizational goals and strategies. This implies that suppliers and the way organizations relate to them are a major determinant of organizational success.

Differences do exist between public and private procurement, between purchasing for service organizations, manufacturers, retailers, distributors, and resource processors; between supplying projects, research and development, job shops, and small and large organizations across a host of industries, applications, and needs. Nevertheless, research has shown much commonality in the acquisition process and its management.

These four textbooks, therefore, cover the common ground of the purchasing field. They parallel the National Association of Purchasing Management (NAPM) Certification Program leading to the C.P.M. designation. They also provide a sound, up-to-date perspective on the purchasing field for those who may not be interested in the C.P.M. designation.

The textbooks are organized into the following four topics:

1. *Value-Driven Purchasing: Managing the Key Steps in the Acquisition Process*
2. *Managing Purchasing: Making the Supply Team Work*
3. *Value-Focused Supply Management: Getting the Most Out of the Supply Function*
4. *Purchasing for Bottom Line Impact: Improving the Organization Through Strategic Procurement*

Volume I, *Value-Driven Purchasing: Managing the Key Steps In the Acquisition Process,* focuses on the standard acquisition process and its major steps, ranging from need recognition and purchase requests to supplier solicitation and analysis, negotiation, and contract execution, implementation, and administration.

Volume II, *Managing Purchasing: Making the Supply Team Work,* focuses on the administrative aspects of the purchasing department, including the development of goals and objectives, maintenance of files and records, budgeting, and evaluating performance. It also discusses the personnel issues of the function: organization, supervision and delegation of work, evaluating staff performance, training staff, and performance difficulties.

Volume III, *Value-Focused Supply Management: Getting the Most Out of the Supply Function,* commences with identifying material flow activities and decisions, including transportation, packaging requirements, receiving, and interior materials handling. It goes on to cover inventory management and concludes with supply activities such as standardization, cost reduction, and material requirements planning.

Volume IV, *Purchasing for Bottom Line Impact: Improving the Organization Through Strategic Procurement,* begins with purchasing strategies and forecasting. This is followed by internal and external relationships, computerization, and environmental issues.

It is a unique pleasure to edit a series of textbooks like these with a fine group of authors who are thoroughly familiar with the theory and practice of supply management.

Michiel R. Leenders
Series Editor

ACKNOWLEDGMENTS

This textbook, like the other three in this series, was based on the sixth edition of the C.P.M. Study Guide. This 1992 NAPM publication, which had been updated by a 1994 supplement, was intended to assist those preparing for the Certified Purchasing Manager (C.P.M.) examinations. The Guide and its predecessors represented the collective work of a large number of purchasing academics and professionals who acted as editors, authors, and reviewers. For the sixth edition these tasks fell to Eugene Muller and Donald W. Dobler, editors, and Harry Robert Page and Eberhard E. Scheuing, consulting editors. Eugene Muller edited the supplement.

The authors for the sixth edition and supplement were:

Prabir K. Bagchi
Judith A. Baranoski
Lee Buddress
Joseph R. Carter
Joseph L. Cavinato
Michael J. Dunleavy
Donald J. Fesko
Barbara B. Friedman
Henry F. Garcia
Larry C. Giunipero
LeRoy H. Graw
Mary Lu Harding
Earl Hawkes
Carla S. Lallatin
H. Ervin Lewis
Charles J. McDonald
Paul K. Moffat
Norbert J. Ore
Alan Raedels
Merle W. Roberts
W.J. "Jack" Wagner
Rene A. Yates

The editorial review board included:

William F. Armstrong
John D. Cologna
Barbara Donnelly
Jack Livingston
Earl F. Pritchard
Karen Swinehart
Marvin C. Sanders
Archie J. Titzman
Earl Whitman

About 53 pages of their work has been used verbatim in this text. It has not been quoted in the traditional academic sense because it has been intermixed so thoroughly as to make clean separation almost impossible.

The same approach has traditionally been used in the preparation of the Guide itself. We made the choice that, if no improvement was possible on what had previously been written, there was no point in rewriting it. We are, therefore, most grateful to the fine work of all C.P.M. Guide contributors over the years.

New ideas in other management areas, as well as procurement, continue to expand the body of professional knowledge in our field. Thus, every attempt has been made to build on the solid foundation laid by many others in the C.P.M. Study Guides and to bring it into relevance for today's environment. This has meant substantial reorganization and additions and a complete format change.

A series necessarily is a collaborative effort. A number of people played a special role with respect to this series. Paul Novak at NAPM acted as the association anchor; Kathleen Little at NAPM served as the ultimate library resource. Jean Geracie at Irwin Professional Publishing got the project started and then passed the baton to Cynthia Zigmund for the next phases.

In addition, a text of this type can only come to fruition based on the efforts of many people. Like most organizations, making this supply chain of professionals work is critical to adding value. In the case of *Managing/Purchasing: Making the Supply Team Work,* we are deeply indebted to a host of accomplices. Marg Reffel, an Administrative Assistant at the Western Business School and Barbara Randa, Learning Resources Center supervisor at Cuyahoga Community College were instrumental in typing this manuscript and providing graphics support. Dominic Gee (Scudder, Stevens and Clark) and Susan Kamauff (KamComp Associates) edited early manuscripts. Jim Simpson (Western Business School) painstakingly edited the manuscript in its early form and contributed throughout its development, especially in the creation of Key Points for each chapter. Steve Golliher, Manager, International Sourcing, and Nolan Taylor, Manager, Sourcing Administration, of Thomson Consumer Electronics reviewed the manuscript for accuracy and practical relevance and contributed to its evolution, and Pat Keating, Vice President, Worldwide Sourcing, Thomson Consumer Electronics, set the example and provided many insightful ideas and comments on how to really make the supply team work.

In terms of content, we gratefully appreciate the assistance of Robert L. Janson, C.P.M., C.P.I.M., Senior Manager, Ernst & Young, a well known purchasing author and advocate in his own right for over 25 years, for his insight and leading edge examples regarding world class sourcing practices. We owe thanks to both Dr. Donald Dobler, emeritus dean of business, Colorado State University and W. Frederick Bartz, director of purchasing for Premier Industrial Corp. for their suggestions that strengthened the original outline of the book. Professor Robert Spekman, Ph.D., Darden Graduate School of Business and Professor Deborah Salmond, Ph.D., University of Baltimore, also contributed significantly to the development of ideas regarding the future of purchasing and the implications for purchasing professionals. Rick Boyle and other professionals from the Center for Advanced Purchasing Studies contributed benchmark studies while Steve Mullaly, a senior consultant for The Victoria Group, provided information on ISO 9000 and quality systems that were adapted for the purchasing arena. Information on EDI was developed with the assistance of the Consortium for Advanced Manufacturing-International (CAM-I) Quality Customer/Quality Supplier (QC/QS) Program whose Program Manager at the time of this writing was John Harcrow. Professor Jim Freeland, Ph.D., Darden Graduate School of Business, provided the information upon which the section on forecasting was developed. In addition, information regarding Value Analysis was adapted from an NAPM tape entitled "Principles of Value Analysis" (Program Aids Library #20) which was produced in 1989, and global sourcing ideas were adopted from the NAPM professional development films "International Procurement Part I: Deciphering Complexities" (Program Aids Library 94) and "International Procurement Part II: Developing the Skills" (Program Aids Library 95) which were produced in 1988 and 1989 respectively.

And finally, we owe a great deal of gratitude to Professor Michiel Leenders, D.B.A., the editor for this series who provided outstanding guidance and insight throughout the process and who patiently led us through the final manuscript development.

We would also like to thank our families, friends, and colleagues for their forbearance during this literary effort. Without their support and assistance we would not have been able to approach this task.

Kenneth H. Killen

John W. Kamauff

CONTENTS

CHAPTER 1

INTRODUCTION AND OVERVIEW

PURPOSE

The effective administration and management of purchasing can contribute significantly to an organization's ability to achieve its goals and strategies. In certain industries such as consumer electronics, it may be vital. This text builds upon the foundation laid in the first book in this series[1] and explores the notion that purchasing, by leveraging the supply chain, can, and should, directly and indirectly, contribute to an organization's long-term competitiveness (Table 1.1).

Purchasing effectively means *making the supply chain work* and often involves challenging existing suppliers, finding alternative sources of supply and creating new linkages both internally and externally. In this context, the *supply chain* is essentially the network of value-added steps that starts with raw materials and eventually contributes to an end product that some customer buys. The concept of the supply chain is equally applicable to manufacturing or service firms; it requires a systems perspective that includes all phases of the product development and delivery cycle and encompasses design, manufacturing, distribution, service and disposal; in essence, all activities from cradle to grave, womb to tomb or lust to dust. Making the supply chain work inevitably implies the development of greater interdependence between members of the supply chain. Unequivocally, the success of any organization may be determined or dramatically enhanced by its ability to maximize the value of its purchases. This text examines the key concepts involved in administering purchasing.

---

[1]Leenders, Michiel R. and Anna E. Flynn, *Value-Driven Purchasing: Key Steps in the Acquisition Process,* Homewood, IL: Irwin Professional Publishing, 1994.

TABLE 1.1
Purchasing's Contribution to Competitive Advantage

TRADITIONAL	FUTURE
Operational	Strategic
Trouble prevention	Opportunity maximization
Bottom line impact	Enhancing the performance of others
Direct	Indirect
Short-term	Long-term
Reactive	Proactive
Evolutionary improvements	Revolutionary improvements

This first chapter provides an introduction and an overview of the book. It explains the importance of purchasing and provides insight into the responsibilities and functions of purchasing administration and management.

Purchasing's Direct Contribution

In terms of direct contribution, for some organizations purchasing is important; in others it is critical. The average North American firm now spends about 60 percent of sales for materials and equipment, according to the *U.S. Bureau of the Census Annual Survey of Manufacturers.* Adding the purchase of services drives this proportion even higher. In contrast, the

FIGURE 1-1
How $1.00 Sales Revenue is Spent by the Average Manufacturing Firm

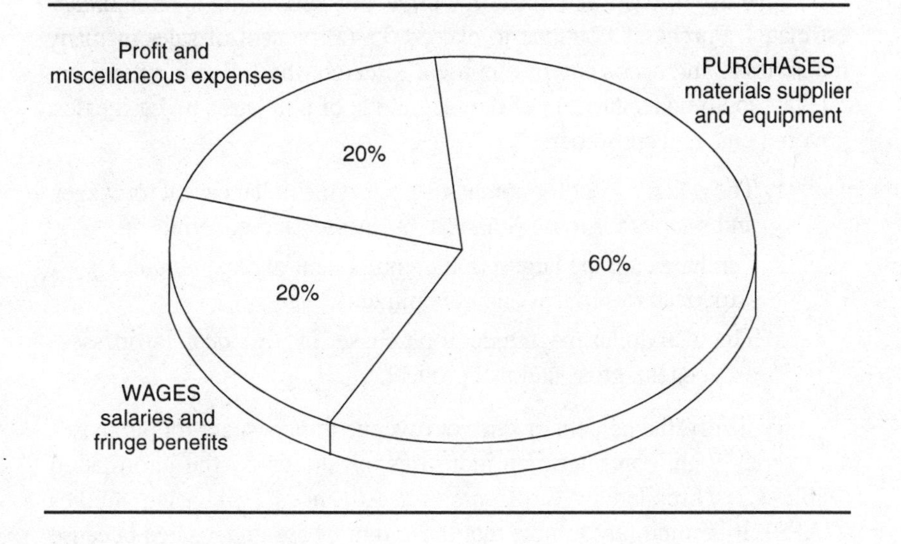

expenditure on wages, salaries, and fringe benefits of all employees amounts to about one third of the materials cost or 20 percent of sales (Figure 1-1). For nonmanufacturing organizations, total purchases may range from a low of 10–20 percent of total revenue in an office environment to close to 90 percent for a wholesaler or distributor. In service organizations such as banks, hospitals, nursing homes, colleges and schools, and various levels of government, purchasing expenditures usually range from 10–25 percent of total revenues.

The fundamental question is, "Why has purchasing (or sourcing, as it is termed in many organizations) become so important?" The simple answer is that value-adding partnerships (whether or not they are collaborative) are becoming more prevalent in operations, particularly with some of the acknowledged industry leaders. Toyota, for example, directly produces only 20 percent of the value of its cars, while General Motors and Ford produce 70 percent and 50 percent, respectively. Chrysler's comeback in the late 1980s was due in part to the creation of a value-adding partnership with its suppliers, distributors, and unions. Chrysler now produces approximately 30 percent of the value of the cars it sells. Many industry observers attribute Ford's recent gains over General

Motors during the 1980s to Ford's aggressive moves to form partnerships with suppliers.[2]

Overall, the cost and economic impact of purchasing is often underestimated. Purchases continue to exceed 55–60 percent of sales in many industries.[3] And according to empirical research, the following items are relevant to an understanding of the magnitude of purchases by businesses, governments and consumers[4]:

1. The dollars spent for purchases exceed the dollars spent for wages and salaries for most American businesses and governments.
2. Purchases are the largest dollar component of expenditures by American businesses and governments.
3. The total dollar magnitude of purchases by American businesses exceeds the gross national product.

Typically, the percent of sales or revenue dollars spent for purchases also varies within and between industries, as shown by the information (Table 1.2) compiled by the Center for Advanced Purchasing Studies (CAPS). It is important to note that these figures are understated because they include only purchases made by the purchasing department, not purchases made by others. In most organizations, a minimum of 10–20 percent should be added to cover such purchases as consulting, insurance, or services, as well as computers and other related equipment. Most people in an organization generally have a healthier respect for the role of purchasing when they are confronted with the magnitude of costs for which purchasing is directly responsible, so it is important to understand and to share this information, particularly with internal customers.

---

[2]Johnston, Russell and Paul R. Lawrence, "Beyond Vertical Integration--the Rise of the Value-Adding Partnership," in *The New Manufacturing, Harvard Business Review* paperback No. 90080, 1991, p. 119.

[3]For additional information on the ratio of purchased materials to sales in specific industries, please see Ansari, A. and Batoul Modaress, "The Potential Benefits of Just-in-Time Purchasing for U.S. Manufacturing," *Production and Inventory Management,* Second Quarter, 1987, pp. 31-35; Leenders, Michiel R. and Harold E. Fearon, *Purchasing and Materials Management,* Tenth Edition, Homewood, IL: Irwin, 1993; and Anderson, Paul, Steve Griffiths and Tim Laseter, "Strategic Sourcing: A Competitive Imperative," Viewpoint, Booz-Allen & Hamilton, 1993.

[4]Heberling, Lt. Col. Michael E., Joseph R. Carter and John H. Hoagland, "An Investigation of Purchases by American Businesses and Governments," *International Journal of Purchases and Materials Management,* Fall 1992, Volume 28, Number 4, p. 44.

TABLE 1.2
Total Purchasing $ as a % of Corporate Sales $

INDUSTRY (DATE)	TOTAL PURCHASING DOLLARS SALES $ (%)	
	AVERAGE	RANGE
U.S. Computer & Telecommunications Equipment (1991)	27	7 - 58
U.S. Computer & Telecommunications Equipment (1993)	34	7 - 55
U.S. Semiconductor Industry (1992)	34	9 - 62
U.S. Transportation Industry (1991)	26	Airlines: 5 - 43.5
		Motor carriers: .26 - 10
		Railroads: 11 - 45
U.S. Aerospace/Defense Contracting Industry (1991)	36	Airframe: 24 - 56
		Electronics: 14 - 57
		Propulsion/Equipment: 18 - 61
		Space: 18 - 38
U.S. Aerospace/Defense Contracting Industry (1993)	36	Airframe: 3 - 94
		Electronics: 8 - 47
		Propulsion: 11 - 48
		Equipment: 26 - 46
		Space: 14 - 74
Electronics	37	22 - 53

The second reason that purchasing is important in terms of direct effects is its profit contribution potential. In assessing this direct contribution, one can assume an average of 5 percent of the sales dollar as profit, although depending on the industry, the year, and the firm itself, this figure can swing dramatically. By increasing sales by $100, $5 in after-tax profits is added. If, on the other hand, sourcing activities lower purchasing cost by $100, the after-tax profit is $50 (using a tax rate of 50 percent), or an amount 10 times greater than the $100 sales increase. Compounded by the realization that most companies have greater control over their costs than their sales, a few people in purchasing often account for a large amount of spending. They therefore have a significant opportunity to reduce cost and thereby increase profit! And in addition to lowering initial purchased prices, purchasing professionals can have a significant impact on quality, innovation, and environmental considerations brought into the supply chain and ultimately delivered to the end customer.

Most companies try to increase profits by reducing labor costs and increasing sales, while at the same time overlooking a significant opportunity in the form of supply chain management. Research has shown that savings in the 5 to 30 percent range are not uncommon when buyers work closely with suppliers to meet ambitious supply objectives,[5] and that "without access to world-class suppliers, a firm will find it increasingly difficult to meet the demands of its sophisticated customers."[6]

Indirect Effects

In addition to the above reasons, Leenders, Fearon, and England cite the following roles that purchasing plays: (1) combating inflation by resisting unwarranted price hikes; (2) significantly reducing dollar investment in materials inventory through better planning and vendor selection; (3) raising the quality level of purchased materials and parts inputs so that the quality and consistency of end product/service outputs can be improved; (4) reducing the materials segment of cost-of-goods sold; and (5) effecting

---

[5]Leenders, Michiel R. and David L. Blenkhorn, *Reverse Marketing: The New Buyer-Supplier Relationship.* New York: The Free Press, 1988.

[6]Monczka, Robert M. and Robert J. Trent, "Worldwide Sourcing: Assessment and Execution," *International Journal of Purchases and Materials Management,* Fall 1992, Volume 28, Number 4, p. 14.

product and process improvements through encouraging and facilitating open communication between buyer and seller, to the mutual benefit of both parties.[7]

In fact, sourcing contributes to cost control as well as to improving overall performance in quality, dependability, flexibility, and innovation.[8] Strategic sourcing is simply the overall process of creating a value-adding (or optimal) mix of supply relationships to provide a competitive advantage. Because materials typically account for more than 50 percent of the product cost, it follows that from a cost perspective alone, effective sourcing can be a precursor to successful new product development. In addition to providing an enhanced opportunity to control costs as well as the prices of its suppliers, many manufacturers have demonstrated the huge value in "product performance and functionality, quality and cycle response times. . . that can be created through a fundamentally different approach to their supply base."[9]

Suppliers are becoming involved in product development earlier (both from a historical and development perspective) than ever because the procurement organization provides access to information it gleans in scanning the vendor base for emerging technologies. In fact, "The greatest contribution potential for supply stems from early involvement in the development of new projects, products, services, and organizational strategies."[10] Garvin argues that in his study of the room air conditioning industry, the clearest indication of purchasing's role at superior performers was its role in New Product Development (NPD). "At the best and better plants, members of the purchasing department were expected to attend design reviews and to keep suppliers informed of new demands. Neither practice was observed at plants with fair or poor quality."[11]

According to Pittiglio Rabin Todd & McGrath (PRTM), a well-developed, well-managed supplier base can yield savings in typical cycle times

---

[7] Leenders, Michiel R., Harold E. Fearon and Wilbur B. England, *Purchasing and Materials Management,* Tenth Edition, Homewood, IL: Irwin, 1993, pp. 2-3. For additional discussion on the importance of a purchasing dollar, see pp. 5-9.

[8] Hayes, Robert H. and Steven C. Wheelwright, *Restoring Our Competitive Edge: Competing Through Manufacturing.* New York: Wiley & Sons, 1984.

[9] Anderson et al., p. 2.

[10] Leenders and Blenkhorn, p. 7.

[11] Garvin, David A., *Managing Quality: The Strategic and Competitive Edge.* New York: The Free Press, 1988, p. 143.

for new product development of 10-20 percent. Improvements in the management of the supplier base have yielded additional ongoing savings of 25 percent to 50 percent in component lead time and 10 percent to 15 percent of annual expense for purchased materials (Kalp, 1991). And in a recent Booz-Allen & Hamilton broad-based survey of the current practices of U.S. manufacturers, firms that employed significantly different sourcing practices consistently performed higher both within industry and across industry groups in inventory turns, incoming material quality, and material inflation (Table 1.3). Finally, in the 1992 Management Roundtable Survey, supply chain management was the key organizational capability most often correlated with being more effective than competitors in terms of the impact on market share.

From these efforts and numerous other anecdotal examples, it is evident that the role of the supply chain in product development is extremely relevant to operations, yet recent empirical research yields some disturbing findings. The 1990 Manufacturing Futures Survey reveals that more than half the respondents begin their process engineering activities (including purchasing and sourcing) after most of the product engineering is finished. They conclude that these actions result in longer product development lead time and missed opportunities in rapidly changing markets.[12] Coupled with the notion that communications with international suppliers, allies, technology providers, and partners are also valid concerns, rigorous empirical research into the robustness of NPD sourcing strategies is warranted.

An additional major role that purchasing fulfills is the task of managing an organization's outside factories—supply management. This recognizes that an organization's success depends, to a degree, on a successful relationship with its outside suppliers. It is one of purchasing's main jobs to cultivate these relationships and to ensure that they continue to be successful. What is meant by a successful supplier relationship and supplier management is discussed later, although this book's main purpose is to describe the principles of good purchasing administration and management.

---

[12]Miller, Jeffrey G. and Jay S. Kim, "Executive Summary," Report on the 1990 International Manufacturing Futures Survey, Boston: Boston University, 1991, p. 14.

TABLE 1.3
Sourcing Performance by Industry Group Summary of Survey Responses

	INVENTORY TURNS			MATERIAL ACCEPTANCE RATE			MATERIAL INFLATION		
INDUSTRY (RESPONDENTS)	WORST	MED	BEST	WORST	MED	BEST	WORST	MED	BEST
Aerospace & Defense (4)	3	7	11	95.0%	97.0%	99.0%	20%	3.0%	-10%
Automotive (11)	2	12	60	96.0%	98.0%	99.97%	3%	1%	-2%
Computers & Electronics (9)	3	5	12	94.0%	98.0%	99.98%	4%	0.5%	-10%
Consumer goods (7)	1	12	75	80%	97.5%	99.0%	6%	1.3%	-25%
Heavy equipment & construction (8)	12	16	30	98%	98.5%	99.98%	4%	3%	1.8%
Industrial products (8)	3	7	50	70%	97.5%	100%	3%	0.7%	-10%
Raw & Basic materials (8)	4	11	120	95.0%	98.0%	100%	4%	0.1%	-3%
Other (6)	3	12	50	90%	98%	99%	5%	3.5%	3%

Source: Paul Anderson, Steve Griffiths, and Tim Laseter, "Strategic Sourcing: A Competitive Imperative," *Viewpoint*, Booz-Allen & Hamilton, 1993, p. 3, reprinted with permission.

THE FIELDS OF PURCHASING ADMINISTRATION AND MANAGEMENT

What is Administration?

According to the *American Heritage Dictionary, administration* is derived from Latin and means "to serve." In this book, *administration* denotes *paperwork*—or at least what used to be called paperwork. Actually, some or most of it may be in computer memory or on diskettes, microfiche, and microfilm. In other words, administration is defined as overseeing purchasing documentation in the context of improving buyer performance.

What is Management?

Originally, *managing* meant the training or handling of horses. (The word *manage* comes from the Latin word *manus,* meaning "hand.") Later, the term was applied to the handling of weapons and vehicles. Today it often applies to managing people or managing a business. From a practical perspective, management, then, is the process of getting things done with and through people, and "managing is a task or activity—a *process*—requiring the performance of several *functions* by individuals possessing a set of *skills.*"[13]

This overview serves two purposes: (1) to provide a condensed summary of what purchasing managers should do and (2) to set the stage for a discussion in greater detail in later chapters of the six functions of purchasing management.

What is Purchasing Management?

Purchasing management is a process of planning, organizing, directing, motivating, controlling, and evaluating purchasing to attain organizational goals. Purchasing management thus entails a set of six interdependent activities that all purchasing professionals should undertake to make the supply chain work. Purchasing management should be goal-oriented and serve as a means to achieve specified results, with purchasing managers contributing to these goals with and through other purchasing professionals.

---

[13]Badawy, M.K., *Developing Managerial Skills in Engineers and Scientists: Succeeding as a Technical Manager,* New York: Van Nostrand, 1982, pp. 4-5.

Purchasing Planning

Planning can be defined as a process of deciding in advance what is to be done, who is to do it, how and when it is to be done, and how well it is to be done. Proper planning in a purchasing context starts with determining the short- and long-term operational and strategic needs. A thorough understanding of current and future markets is required. It is also important to resolve who in the organization will be responsible for the various requirements and when purchases will be made. Additionally, a significant aspect of the planning process is the explicit delineation of how the various purchasing strategies will be executed to achieve maximum value for the organization.

Operational planning ensures that the right materials or services of the right quality, quantity, and price arrive at the right time in the right place. Strategic purchasing planning assures that the organizational goals and strategies are properly supported by the supply function. It is also important that organizational goals and strategies effectively incorporate purchasing and supply chain opportunities.

Organizing for Purchasing

Organizing is the work of providing in advance those things needed to carry out the plan, and it should reflect the main elements of the plan. If purchasing planning is not effectively accomplished or is nonexistent, then the organizing cannot truly contribute to the firm. Once a good plan has been constructed, other factors to consider in organizing include:

1. Making the right people available.
2. Ensuring that these people know their roles in the organization and how those jobs relate to others.
3. Confirming that these people know what part of the plan is their responsibility.
4. Guaranteeing that they are properly trained to carry out their part of the plan.
5. Providing the requisite support resources needed at the right time and in the right place to enact the plan.

In today's world of the "new competition," reengineering, downsizing, continuous improvement, employee empowerment, and focus on value-added activities, the challenge of organizing the supply chain effectively is substantial. Many of the traditional norms of department structure and mandate have been found wanting. Also, the competencies of supply personnel who can contribute well under the new organizational perspectives have to be different from the qualifications of the traditional purchaser steeped in policies and procedures that no longer support organizational initiatives.

Directing the Purchasing Department

Directing entails the use of communications and leadership to guide the performance of one's subordinates toward achieving the organization's plans. In the supply management context, directing tasks requires individuals who themselves are convinced of the necessity for change and are confident their plans will yield improvements. A relatively new phenomenon is the emergence of a new breed of purchasing managers who have a non-supply background, such as marketing, sales, engineering, manufacturing or finance. Obviously, in this situation the need for a new direction is underscored by top management, and it is a rather telling argument that no one in the previous supply organization was deemed capable enough to take over the reins. The depth of knowledge traditionally gained in the negotiation trenches over long periods of time, which previously served senior purchasing professionals so well, is no longer sufficient, given the demands of today's marketplace. It is the contention of this book and the others in this series that those individuals aspiring to management ranks in purchasing also need breadth and exposure to other disciplines if they are to tackle the vast challenges facing our profession (and most organizations) at this time.

Motivating Purchasing

Motivation is the process of channeling a person's inner drives so that he or she wants to accomplish the goals of the organization. Motivation is not something that you do to employees to make them work. Rather, employee motivation is achieved by providing the proper psychological work environment.

The difficulty in motivating within purchasing is in getting existing staff to embrace the new challenges without sub-optimizing their efforts. "Am I adding value to the organization by doing this?" is a perilous question for anyone in any position at any time, but it is a question that has to be revisited time and time again. For many purchasers the question may well reveal that a significant portion of their lives has been spent on activities with low added value. The upside is that for those purchasing professionals who are excited about the prospects of change and are willing to respond to (if not anticipate) the evolving role of supply management, their contributions to the organization can be magnified and their individual working lives can become more meaningful.

Controlling Purchasing

According to Henri Fayol, the French father of management, "Control consists in verifying whether everything occurs in conformity with the plan adopted, the instructions issued, and principles established." *Control* is the process of checking up on the actual progress of a plan. Actual conformance to a plan requires supervision, buy-in, and feedback. The effectiveness of the plan cannot be established if conformance does not exist in the first place. Ideal conformance derives from individual and collective conviction regarding the soundness of the plan, rather than from forced acceptance under duress. Appropriate consultation and involvement by purchasing personnel (as well as internal customers, other functions, and suppliers) in the development of the plan substantially assists in conformance and often mitigates the need for heavy-handed control mechanisms.

Evaluating Purchasing

Evaluation is the final step in the purchasing management cycle before embarking on the planning activities for the next cycle. Evaluation has a dual aspect in this context. Differing from controlling, which validates conformance to plan, evaluation establishes whether the plan is still appropriate under inevitably changing environmental conditions. Have internal and/or external conditions changed? Did the plan take into account all relevant factors? Evaluation also looks forward and addresses the issue of whether what we are doing now will be appropriate for the periods ahead. The status quo will be maintained only if evaluation proves it to be the best action for the future. Regular evaluation is, therefore, essential to ensure that the organization continues to function effectively.

It should be obvious by now that planning, organizing, directing, motivating, controlling, and evaluating embodies a continuous cycle of management activities. As shown in Figure 1-2, this cycle is repetitive. For

FIGURE 1-2
The Purchasing Management Cycle

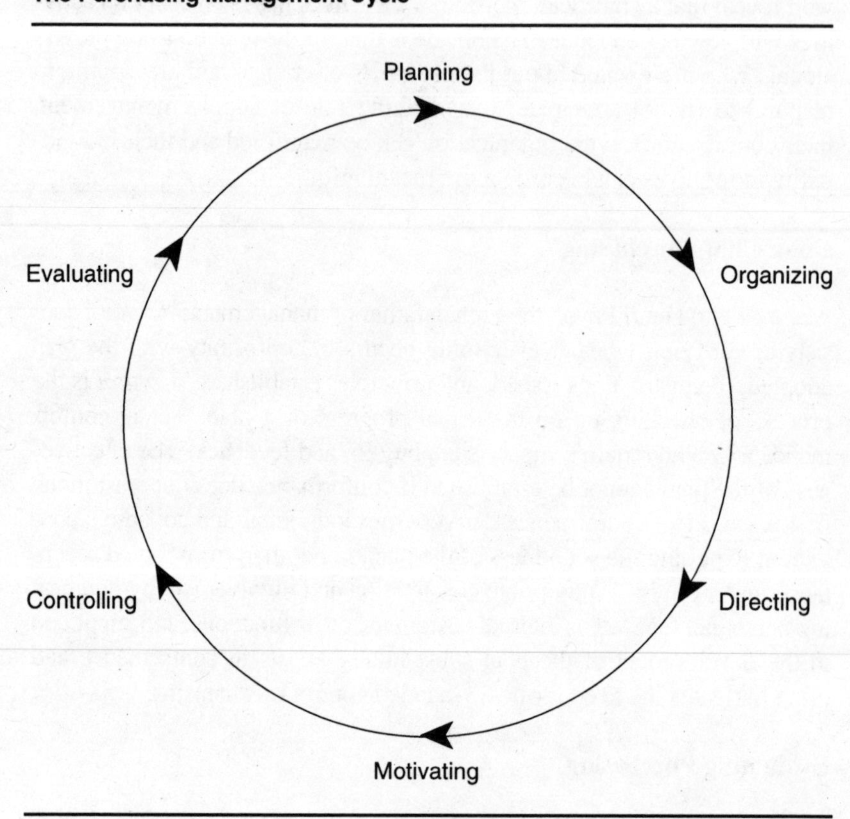

purchasing, evaluating results against operational and strategic objectives represents a significant challenge. Typical operational measures include quality, delivery, quantity, and price. Strategic measures are often more difficult to quantify and may include process measures and more substantive attributes such as innovation, time-to-market, cycle time, and responsiveness.

MANAGING PURCHASING

This book's objective is to take the reader through the key functions of purchasing management and its administration. These key functions are planning, organizing, directing, motivating, controlling, and evaluating. Chapters

2 and 3 focus on administrative paperwork and systems, and present basic concepts of forms design and record maintenance. Chapter 4 begins the discussion of key management functions and introduces purchasing planning. It discusses forecasting, explains the importance of planning, and describes the types, techniques, and methods used to develop goals and objectives.

Chapter 5 explores financial planning, budgeting, and cost savings. Chapter 6 investigates organizational principles, design, and structure, and discusses the advantages and disadvantages of centralized versus decentralized purchasing. Chapter 7 continues the examination of organizing as it applies to staffing and concentrates on such topics as selection, promotion, dismissal, and training. Chapter 8 is devoted to directing, and it emphasizes leadership and communications. Chapter 9 summarizes current motivation research and addresses how it can be applied in a typical supply management environment.

Chapter 10 investigates control, staff performance assessment, the reasons control is needed, how to control, and the explicit relationship between control and the other management functions, especially evaluation. Chapter 11 concludes with a forecast of what is ahead for purchasing in the "new competition" as it applies specifically to organization, international business, and the role of the purchasing professional.

KEY POINTS

1. The average manufacturing firm now spends about 60 percent of its sales dollar for materials and equipment.
2. Service organizations ordinarily spend in the range of 10–25 percent of total revenues for purchases.
3. In manufacturing organizations, the profit contribution, through cost reduction, is 8 to 10 times greater for purchasing than for sales.
4. Purchasing can contribute to the overall success of its organization by combating inflation (keeping the organization competitive), helping to make large reduction inventory investment, raising quality levels, lowering the cost of goods, and facilitating product and process improvements.
5. Purchasing administration is defined as overseeing the process of documentation of the department.
6. Good purchasing management is a process of planning, organizing,

directing, motivating, controlling, and evaluating to achieve organizational goals.

7. Planning is a process of deciding in advance what is to be done, who is to do it, how it is to be done, when it is to be done, and how well it is to be done.
8. Organizing is providing in advance those things needed to carry out the plan.
9. Directing is the use of communications and leadership to guide the performance of one's subordinates toward the achievement of the organization's plans.
10. Motivation is the process of channeling a person's inner drives so that he or she wants to accomplish the goals of the organization.
11. Controlling is the process of verifying the actual progress of a plan.
12. Evaluation is the process of confirming a plan's effectiveness and its future applicability for guiding the organization.

REFERENCES AND RECOMMENDED READINGS

Leenders, Michiel R., Harold E. Fearon, and Wilbur B. England, *Purchasing and Materials Management,* Ninth Edition, Homewood, IL: Irwin, 1989, pp. 2 and 3.

CHAPTER 2

THE ADMINISTRATIVE PAPERWORK

PAPER AND THE COMPUTER

This chapter and the next examine the records that are ordinarily kept in purchasing, discuss how and why to use the existing administrative systems, and offer suggestions for developing effective ways to administer the purchasing system to leverage the supply chain. Purchasing paperwork consumes a purchasing professional's available time and threatens the fundamental ability of purchasing to add value to the organization. According to Jerry Brown, former governor of California, "The volume of paper expands to fill the available briefcase."[1] If they become preoccupied with mundane paperwork, purchasing professionals become record keepers and order takers. Purchasing administration (which includes systems management, filing, data entry, internal reporting, external reporting, and staff and department evaluation), like all other purchasing tasks, must yield both direct and indirect contributions if purchasing is to provide a distinctive competence for the organization.

According to Mike Roberts of RPM Associates in a recent article in *Purchasing,* "Activity analysis reveals that Operating and Administration activities represent 80 percent of the time, effort, and cost of a typical procurement department while activities devoted to Strategic Development and Supplier Relations generate the more significant long-term results."[2] Table 2.1 shows this relationship between the cost of purchasing activities and the long-term business impact.

In many organizations, existing administrative paperwork often serves merely to document a chain of events or to provide a logistical paper trail.

---

[1]Boone, Louis E., *Quotable Business,* New York: Random House, 1992, p. 138.

[2]Porter, Anne Millen, "Tying Down Total Costs," *Purchasing,* October 21, 1993, p. 41.

TABLE 2.1
Typical Procurement Activities

ACTIVITIES	COST OF ACTIVITIES (%)	LONG-TERM BUSINESS IMPACT (%)
Strategic development	5	40
Supplier relations	15	30
Administrative	25	5
Operating	55	25

Leading edge purchasing organizations need to transform this administrative function into a value-adding process by reducing, eliminating, or combining steps wherever possible. Merely automating only speeds the existing processes, when it is often necessary to reengineer the administrative paperwork itself. This chapter will review traditional purchasing administration and will suggest alternatives for change.

Even though many purchasing departments have become more computerized, it is still appropriate to speak of paperwork. Almost every report or form found on a computer has at some point been done without a computer. For the most part, purchasing organizations have merely automated what existed previously. Using a computer for record keeping and report generation usually changes little about the materials as to how and why they are needed and used. Automation just makes it easier to get information faster with less effort, which may or may not be useful, as people often tend to hoard unimportant information.

To understand computerized purchasing, one must first understand the documents and the documentation systems. In this book, we discuss documents as if they were pieces of paper, realizing that these can and are being computerized. In fact, by very early in the next century, almost all buying and selling will be computerized—at least the records will be.

ADMINISTRATION VERSUS THE FUNCTIONS OF MANAGEMENT

Administrative tasks cannot be separated from planning, organizing, directing, motivating, controlling, and evaluating. In fact, purchasing administrative tasks are interwoven throughout the six management functions. For example, restrictive paperwork requirements often adversely affect motivation. Record keeping is typically nonproductive (although it admittedly serves important purposes such as providing a record of a supplier's past performance); it does not create a product or service and thereby contributes nothing directly to output. Therefore, documentation should be kept to a minimum, although it is important to recognize that training needs are often inversely proportional to documentation requirements, as shown in Figure 2-1. Many Japanese companies do not use purchase orders, receivers, or invoices. They rely on less formal structures; often a handshake replaces the purchase order, and they compensate based on the build schedule. For example, if Toyota builds 100 automobiles they know they must have received 100 good radiators from the supplier, so they remit payment for the agreed-upon price. The automobiles rolling out the

FIGURE 2-1
Documentation versus Training

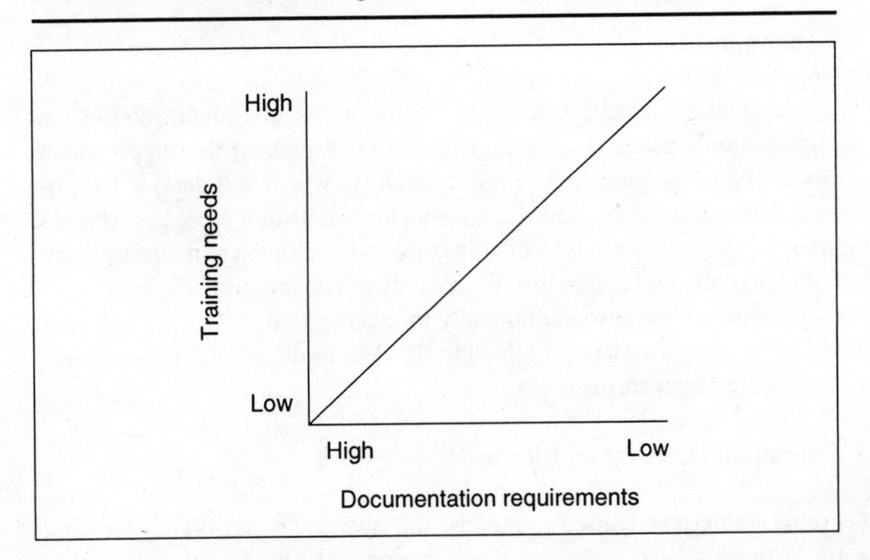

door are the only proof they need that the parts were received and are functioning properly. By establishing an effective relationship with this supplier, they can eliminate nonessential work; no inspection or documentation is necessary. The relationship builds on trust, and the procedures (and problem-solving, if necessary) are determined jointly. Thus, the value-added up-front planning enables them to eliminate the nonvalue-added paperwork later.

REASONS FOR FORMS

Despite this desire to eliminate paperwork, a fundamental task for purchasing administration is to design and administer the use of operational forms. Communications, operational control, records, and consistency of approach are the most significant reasons for using forms. The first book in this series describes the key steps and the accompanying forms used in the value-driven acquisition process. Table 2.2 reviews these basic process stages in a traditional purchasing system and includes the typical forms necessary for each step. The Thomson Consumer Electronics purchasing process, a highly effective annual contracting system which is used to acquire $4 billion in components and finished goods annually, is shown in Table 2.3 to provide an example of when these forms might be used.

Communications

As depicted in Table 2.3, for most organizations, communication of the organization's needs is a fundamental facet of making the supply chain work. The communication of information (in words and data) is the primary reason forms are used. Throughout the acquisition process, forms are often the most efficient way of collecting and communicating information both internally and externally. To make them better communication tools, they should be reviewed continually to make sure that they are still providing useful information. Allowing the data to dictate the requirements can lead to better forms.

Operational Control and Records

Forms are frequently used to exercise management control. They are used for activities such as controlling transactions, funds, and schedules.

TABLE 2.2
The Acquisition Process and Most Commonly Used Forms

STEP	FORM USED
Determining needs	Material requirements Planning (MRP), forecast, bill of materials (BOM), planning documents, or inventory replenishment form
Communicating needs	Purchase requisition (PR)
Identifying potential sources	Past supplier records, directories ries, recommendations from users or designers, request for information (RFI)
Soliciting and evaluating bids & proposals	Invitation to bid, request for quotation (RFQ), RFI, request for proposal (RFP), bid analysis or bid summary
Preparing the purchase order (PO)	PO, contract, blanket order, systems contract, blank check order, telephone order, credit card or release against contract
Following up & expediting	Follow-up or expediting form
Receipt & inspection	Receiving report, weigh bill, shipping document, excise and customs document, inspection report, in-process report, stock-ing report or inventory records
Clearing the invoice & payment	Invoice, payment, automatic payment options (funds transfer)
Maintaining records & determining cost standards	Supplier records, PO records, commitment records or stock keeping records

TABLE 2.3
Thomson Consumer Electronics Acquisition Process

Thomson Contracting System (TCS) Activities		*Date*
Commodity managers & team members appointed		Feb 20
First cut bidder's list issued to locations		Apr 19
Bidder's list comments returned to Indianapolis		May 4
RFQ requirements complete		May 19
Review final bidder's list/commodity strategy (Indianapolis)		Jun 4-9
Kick-off meeting (Indianapolis)		Jun 7
RFQs mailed to suppliers		Jun 18
Intent to bid confirmation		Jul 1
Bids due		Jul 22
Bid evaluations	North America	Aug 8-17
	Taiwan	Aug 20-23
	Asia	Aug 28-29
	Europe	Sep 10-14
Negotiations		Sep/Oct
Business plan commodity factors report		Oct 1
Complete negotiations		Oct 30
Wrap-up meetings	North America	Nov 8
	Asia	Nov 9
	Europe	Dec 12
Cost standards complete		Nov 12

Another reason for using forms is to provide a record of activities—a history of what has been done. These records are needed to research information, prove that actions did or did not occur, provide audit trails, etc., and they typically fall into two classes, short and long term. Short-term working records simplify the work of the department on a daily basis. For example, during the buying cycle, open requisitions, bid documents, and copies of open purchase orders are nearby so the buyer can refer to them until the buying cycle has been completed. After the buying cycle is complete, the records become part of the long-term history of the department. Probably 80 percent of these records will never be consulted again, but rarely does an organization know which 20 percent will be needed in the future, so most records are maintained. Furthermore, there are usually legal requirements to maintain certain records for a minimum period of time.

Consistency of Approach

Forms allow tasks to be accomplished consistently and ensure that all pertinent information is included. In addition, forms allow greater efficiency (when preparing requisite information from scratch) because the same sequence and placement of information take place every time the form is used. The individuals using forms can also glean information more easily than if it were transmitted in many different formats and styles. Moreover, all receivers of forms get identical information, which adds an ethical dimension whereby all suppliers have access to the same information for bidding purposes.

Frequently when purchasing consults long-term records, it is to respond to requests for information from outside the department, such as engineering or accounting, or for the department's own research. These long-term records provide a paper trail for accounting, internal and external auditing, and for legal or tax purposes. Additionally, records are sometimes needed for EEOC, EPA, OSHA, or other types of regulatory requirements at the federal, state, and local levels.

TYPES OF FORMS USED IN PURCHASING

The purpose of this section is to provide a list of common forms used in purchasing and to explain their use.

Requisition

The acquisition cycle usually starts with the recognition of a need, which is typically translated into a requisition, although occasionally a master schedule or other document may be used. The requisition commonly is an efficient way of communicating the user's needs and becomes purchasing's authority to spend money. The purchasing department should have a list of those people who have the authority to sign requisitions and should not accept requisitions unless signed by an authorized person.

The purchase requisition is commonly used to request that purchasing buy certain items. Various departments forward this form to purchasing, indicating the items they want and where and when they want them delivered, and including the appropriate accounting and approval information. Simply put, the purchase requisition authorizes purchasing to make a buy.

There are two types of requisitions: disposable and traveling. The disposable requisition is filed or discarded after each use. On the other hand, fewer than 10 percent of purchasing organizations typically use the traveling requisition (TR). TRs are more cost efficient than disposable requisitions because they are used for repetitive purchases, and they provide a reliable tracking system for items. The TR is printed on card stock with standard information preprinted on the form (such as the description of the item to be purchased and accounting information), because, depending on the frequency of use for an item, one TR may last five to ten years before it is filled up. This form provides the requisitioners with the advantage of reducing preparation time, since they normally have to enter only the items that do change (e.g., quantity and date). The advantages of the TR are that preparation time is typically reduced and a history of previous expenditures for the item being purchased is immediately available on the form itself. TRs are commonly used for such items as storeroom materials, although they can be adapted for a wide variety of uses.

The TR also has room to list from three to six approved suppliers, and a space is usually provided for the part number and accompanying part description. The remainder of the card is ruled both vertically and horizontally. The following columns are recommended:

1. Requisitioner's initials.
2. Request date.
3. Quantity needed.
4. Date needed.
5. Date purchase order (PO) is issued.
6. PO number.
7. Expected arrival date.
8. Buyer's initials.
9. Date order was delivered.

Although these general recommendations should be followed, the TR should be tailored to fit the needs of the individual organization. Rarely is purchase price included since this information should be kept confidential and shared only with those who need to know (often only purchasing and accounting).

The requisitioner files the TR within his or her respective department, and when a new order is necessary, the date and quantity are added to the existing TR. Subsequently, the requisitioner sends the TR to purchasing. Purchasing completes the remaining columns, except for the *date order arrives* column, and returns it to the requisitioner. To ensure accountability, the buyer's initials are added. The requisitioner checks the information that purchasing entered and files the form in an open order file. If the requisitioner does not concur with purchasing (for example, if the delivery date will be later than needed), he or she must contact purchasing immediately to resolve the discrepancy.

When the requisitioner is notified that the item has been received, the date is entered on the card, and it is filed. If the order received is only a partial delivery, then the TR is kept in the open order file until the delivery is complete. It is a good idea to color code the cards by department or commodity group and to print the columns on the back of the cards so both sides can be used. Management approval for requisitions become more efficient through using "encumbrance style" pre-ordering computerized authorization, which assures that departmental budget funds are available at the time of requisitioning without additional signature reviews.

Bill of Materials

A bill of materials (BOM) is a diagram or record of all the components of an item, the parent-component relationships, and usage quantities.[3] It lists the parts needed to produce a component, intermediate items, subassembly, or complete unit, and it specifically identifies purchased items. Bills of materials are sometimes used instead of requisitions, and they usually originate with the design engineer or from a material requirements planning (MRP) system for use in manufacturing and production planning. The requester typically informs purchasing of the total amount of the product to be produced, and purchasing uses the BOM to determine the quantity of individual parts to buy.

Contracts

The purchase order (PO) is a form used by purchasing to contract with suppliers. It is the most common purchasing contract form because it is

---

[3]Krajewski, Lee J. and Larry P. Ritzman, *Operations Management: Strategy and Analysis,* Reading, MA: Addison-Wesley, 1993, p. 502.

designed to save time and should be preprinted with a well-defined set of standard terms and conditions. Common elements of information on most POs include: name of the supplier, price, quantity, product description, shipping and billing information, delivery date(s), payment terms, and date of order. POs are usually multipart forms, with the number of parts determined by an organization's needs.

A PO is the easiest and the most efficient method to enact a purchase contract, although it is not the only way that such contracts can and are made. The prominent features to include on a PO are:

1. The words ***"PURCHASE ORDER"*** in bold type at the top.
2. ***For whom*** it represents and the organization's name and address, phone number, and facsimile number. (The word *for* is used to denote on whose behalf the agent is acting.)
3. ***Freight terms,*** which should allow enough space to include a statement such as: Free on Board (F.O.B.—meaning that once the cargo is placed on board a vessel, at the place designated in the purchase contract, then the seller is free of obligation for loss or damage), Ex-Works (meaning that the seller must deliver the goods at his or her own premises and provide the buyer with the country of origin documents needed for export and subsequent import), or point of origin (typically, state, city, and county should be listed).
4. ***Payment terms.*** In this context, terms of *1/10/net 30* earns 18 1/4 percent annual interest, whereas the more traditional term of *2/10/net 30* yields 36 1/2 percent annual interest.
5. A ***PO number*** for control purposes. Accounting should issue these numbers at the beginning of the accounting year, and purchasing should be responsible for each.
6. ***Space*** for the quantity and item description.
7. ***Other terms*** of the sale such as: (a) seller warrants that they comply with all local, state, and federal regulations and laws, (b) seller warrants that they have not violated any patents, copyrights, or trademarks, (c) seller—if providing on-site service—must furnish evidence of both paid-up liability protection and Worker's Compensation, (d) order must not be filled in greater quantities than indicated, (e) no charge for packing or drop shipment will be allowed, (f) seller agrees to protect and return, upon buyer's request, all tools, drawings, etc., and (g) all documents involved with this sale must have the buyer's purchase order

number on it. These are typical examples of common terms, but there are others that might be used, so it is wise to consult the legal staff before printing the forms.

8. A ***statement of terms*** on the front of the order should call the reader's attention to the back of the order if there are additional terms and conditions. The statement could be worded: "This order must be accepted only in accordance with the terms and conditions set forth on the front, reverse side, or other ways attached and noted. No other terms and conditions will be accepted if not first approved in writing by the undersigned buyer." Failure to have a statement of this type may cause any terms below the signature line to be disregarded in a court case.
9. A ***signature line*** that has the word *by* before it to clearly show all parties that the signer is an agent, not the principal, in the contract.

In organizations with top purchasing functions, PO contracts are usually complex and specify commodity details, specific item/part release quantity pricing, prenegotiated hourly labor rates, and material costs for customer requirements. Invoices from suppliers and payments to these organizations are determined based on carefully worded POs transmitted by electronic means or by using periodic statements; they often include supplier-furnished price lists on computer disk which are used to electronically match POs to invoices.

The Receiver

Receiving documents (also known as receivers or receiving reports) are used to indicate that a shipment has arrived. The PO and receiver are alike because most of the information contained in the PO has to be duplicated on the receiver. Frequently, the receiving document is one of the designated copies of the PO. Often, the only thing not on the receiver that is usually shown on the PO is the purchase price. Some companies also omit the quantity from the receiver because they want to ensure that the receiving department makes an accurate count. However, it is probably appropriate to provide the quantity because this information is needed by the receiving department to make sure that the entire shipment is received rather than a split order.

However, many companies take advantage of the fact that the purchase order provides information that is almost identical to that needed on the receiver, so most manufacturers of business forms sell a combined PO/receiver form. This saves time and money because it eliminates the need to type redundant information on the receiver. After the PO is typed, the receiver packet is separated and sent to the receiving department.

The arrival of a receiver or receiver packet triggers the need to provide information for planning purposes. By knowing what is expected and when, the receiving department supervisor can plan the department's work better. The ideal situation is to have 100 percent of the items delivered on the day they are due. Planning is difficult when only 65–75 percent of the items arrive when they are needed; thus, many people engage in partial planning rather than not planning at all. Without at least partial planning, the receiving department requires idle (or nonvalue-added) labor awaiting incoming calls. The receiving department also needs to know, for example, where to deliver the incoming shipment and which department should be charged the receiving cost. Since this information is (or should be) contained on the requisition, it can easily be typed on the purchase order/receiver, thereby making it readily available.

Another reason for having the receiver packet prepared as part of the purchase order is that it reduces cost by eliminating duplicate typing. In fact, it eliminates the necessity for ever having a typewriter or word processor in the receiving department. The only things that need to be entered on the receiver packet are the quantity received, the date, the receiver number, and the signature of the receiving clerk. The most common criticism of the combination PO/receiver form is legibility, because some systems require as many as eleven copies.

Inspection Documents

Inspection documents differ from receiving documents in that they are used to record whether the goods conform to the organization's requirements. These forms typically specify the type of inspection used (e.g., 100 percent inspection) and the results of that inspection (e.g., 0.2 percent defective). In many nonmanufacturing organizations, there is no centralized inspection activity, and inspection becomes the responsibility of the receiving clerk. In such cases, the receiving report and the inspection document are one and the same.

Change Orders

Change orders or change notices are forms used to modify or cancel purchase orders that have been issued. Some organizations use special change order forms, while others use their standard PO form with "Change Order" printed or stamped on it. Since the PO is the written evidence of a contract, it is appropriate that all PO changes be in writing. In addition, if a contract has already been formed, the change must be accepted by the other party.

Material Return Forms

This form is used to notify the supplier and in-house departments such as purchasing and inventory control that goods must be returned. The reasons for returning these items may include over-shipments, rejections, or damaged goods.

Computer-Generated Forms

The advantages of computers, as discussed elsewhere in this series, also apply to the generation of forms. Probably the most commonly used computer-generated form in purchasing is the PO. A PO data file can be used to process POs efficiently and to provide numerous reporting and status activities. Almost any form can lend itself to computer applications. Many computer-generated forms are used in a paperless system in which the form itself is viewed and used entirely on a CRT.

Other

Although the previously described forms are those that are most commonly used in purchasing, there are other forms as well. These include forms used for bid analysis, request for quotation, price records, supplier ratings, tracer request, surplus equipment forms, and quality performance records.

In the next chapter we explore how purchasing managers can design administrative systems and use these forms as management tools that contribute both directly and indirectly to creating a competitive advantage for purchasing.

ELEMENTS OF FORM DESIGN AND MANAGEMENT

Available Media

Available media for forms include many possibilities ranging from standard paper to paperless forms. The appropriate media depend on such factors as the form's immediate and future uses, cost, storage, ultimate disposition, and even the image the organization wishes to project.

Organization of Information

The organization of information on a form is important to its successful usage. Form design necessarily constitutes documentation, and when completed, the form becomes a record. One of the key elements to consider is that the information should be kept to the necessary minimum. Superfluous information detracts from finding needed information. Additionally, to make it easy for the user to enter information on the form, many forms allow check marks next to appropriate responses, as opposed to a "fill in the blanks" format. Another important element is to keep the information in a sequence that makes sense; adhering to the logical order of information should be accommodated on the form.

Instruction

If at all possible, instructions for the use of a form should be on the form itself. The instructions need to be concise and clear to all the individuals who will use it.

Forms Control

Forms control is important in any organization, and in larger organizations, this activity is usually centralized. In smaller organizations, it is often performed by each department for that particular department's forms. Forms control involves the key elements of design, user determination, standardization, and appropriate stock levels. An important element of forms control is the periodic review of all forms to ensure uniformity and to reduce obsolescence.

Retention Periods

Often, compromises between cost, convenience, and need determine retention periods. The questions to ask in determining how long to retain forms on file include:

1. How long will the information be needed?
2. What are the legally required periods, if any, for storing the forms?
3. What is the cost of storage?

Privacy Act/Public Information Implications

The Privacy Act places restrictions on access to some types of information, which has important implications in form design. In governmental purchasing, most (if not all) purchasing forms are a matter of public record and are subject to public review. Legal advice is often necessary when dealing with the Privacy Act/public information implications of forms, and, as a minimum, purchasing has an obligation to address these issues.

Cost Implications

The cost of printing a supply of forms is a fraction of the form's true cost. Often, the clerical time needed to process the form can be significant, particularly if the form is poorly designed for the existing typewriters or word processors. Additional costs are incurred for storage and distribution (including copies). But the most significant costs can be associated with slowness of information access, especially if the form is poorly designed. Finally, substantial costs can result from errors in design or form completion.

World-Class Practices

The administration and management of forms are often onerous but seemingly unavoidable tasks for purchasing. Still, purchasing professionals must continue to evolve from a longstanding tradition of being record keepers and order takers. Merely automating the existing processes will not suffice to revolutionize purchasing's role in most organizations. Purchasing administration (which includes filing, data entry, internal reporting, external reporting, and staff and department evaluation), like all other purchasing tasks, must yield both direct and indirect contributions if purchasing is to provide a distinct competitive advantage. Adopting a forms management philosophy based on customer-focused information flows can reduce the amount of nonproductive time an organization spends and can free purchasing professionals for more added-value pursuits.

An example of this move away from clerical tasks is the reduced reliance on individual purchase orders. According to Robert L. Janson, C.P.M., C.P.I.M., senior manager, Ernst & Young, in world-class purchasing, 70 percent of dollars purchased are made through the use of regional contracts, blanket orders, national agreements, or similar arrangements. Additionally, requisitioners communicate 70 percent of the paperwork (e.g., POs or releases against blanket orders) *directly* (often electronically to minimize lead times and to enhance responsiveness) to the respective suppliers. With decentralized direct releasing, the number of POs is dramatically reduced but the number of "releases" increases. Purchasing is not involved in these ongoing activities but continuously performs exception-based checks and balances. In this manner, purchasing contributes added value by creating contracts, not executing them.

As an example, Wilson Sporting Goods in Humboldt, Tennessee, established blanket contracts, thereby reducing materials cost and stabilizing prices on materials for one or more years. Discount payments also resulted in savings of more than $100,000 per year, and the cost of holding inventory was reduced by 67 percent while annual inventory turns were increased.[4]

To make this administrative process work, people and systems from throughout the supply chain must work together. Purchasing professionals must aggressively integrate the paperwork needs of internal customers *and* suppliers, because any nonvalue activities associated with administering the supply chain are detrimental, regardless of where they occur. Adopting a systems perspective that takes a holistic view of the supply chain is a fundamental step toward minimizing the adverse effects of administrative paperwork. The overriding goal should be to reduce, simplify, or eliminate paperwork wherever possible. Only by continually reevaluating and subsequently synchronizing the paperwork requirements of all members of the supply chain can an organization effectively achieve world-class purchasing.

KEY POINTS

1. Forms are used to collect data, communicate information, provide records, and serve as a tool for management control.
2. Using a computer for record keeping and report generation usually changes little about the format of the forms or how and why they are needed and used.
3. Documentation should be kept to a minimum because it is a nonproductive activity.

---

[4]"Being the Best," *NAPM Insights,* April 1993, p. 45.

4. An organization should review its purchasing forms continually to ensure that they are needed and appropriate.
5. The two types of purchase requisitions are disposable and traveling. They express the requisitioner's needs and give purchasing the authority to spend the requisition money. Traveling requisitions are more efficient because they can be reused.
6. Bills of materials list the parts needed to produce a component, subassembly, or complete unit. They serve as a replacement for a requisition.
7. A contract is an agreement between two or more parties that clearly states what each party agrees to do, and it is enforceable by law.
8. The PO should be the most common purchasing contract form because it saves time.
9. Contract forms other than the PO may be used when the contract period exceeds an accounting year, when orders are placed electronically, or in special cases such as capital equipment, construction, and certain services.
10. The receiving documents indicate that a shipment of goods has arrived. They notify departments of the shipment and show whether there is a problem with the shipment.
11. Inspection documents are concerned with assessing the quality of an incoming item.
12. Change orders or change notices are forms used to amend the PO. If a contract exists, the other parties to the agreement must approve the change.

REFERENCES AND RECOMMENDED READINGS

Leenders, Michiel R., Harold E. Fearon, and Wilbur B. England, *Purchasing and Materials Management,* Tenth Edition, Homewood, IL: Irwin, 1993.

CHAPTER 3

THE ADMINISTRATIVE SYSTEM

OVERVIEW

Notwithstanding the significance of paperwork in administering purchasing (as established in the previous chapter), it is necessary to transcend the traditional reliance on existing purchasing administration to make the supply chain work effectively. To effect this transition requires examining the whole purchasing system (from end user to raw materials suppliers) and considering administration as it affects the entire supply chain rather than the purchasing department alone. It is important that this systems perspective is well established prior to investigating the functional responsibilities of purchasing professionals. This chapter attempts to provide this systems outlook using administrative paperwork and tasks as a foundation for subsequent analysis of planning and forecasting.

MANAGERIAL SYSTEMS

The word *system* comes from the Greek language and means "to place together" or "an arrangement of related things" to form a unit. In other words, a system is an orderly way of doing something and can be anything that maps inputs to outputs. Organizations have worked out uniform ways of doing things, and we call them *managerial* systems. Only those systems directly relating causes to effects that have been consciously established by management can be considered true managerial systems. In essence, a management system assures that *planning* is carried out such that *all* staff know what is *expected* and *how* to achieve specified *results.* The system encompasses the technical, administrative, and human factors plans and functions needed to govern operations, and it is conducted based on data and facts.

For example, most organizations have purchasing systems in place that administer their interactions with the supply chain. Such systems typically allow its organization to receive requests to place an order, select suppliers, place the orders, track the orders, and receive the orders. These systems enable purchasing to support the firm's long-term supply needs and to support the firm's production or service delivery systems.

Why Are Managerial Systems Needed?

We would be in bad shape in business or industry, or in any other organization for that matter, if we had to start each day anew as if each person had never done that job before. By using managerial systems to explicitly map actions with results, these systems enable organizations to be more productive (i.e., to derive a higher ratio of output to input) and, therefore, more efficient. Another reason for using managerial systems is that individuals and organizations "learn" these efficient practices, which subsequently become ingrained and make it unnecessary to "reinvent the wheel" every time a task

FIGURE 3-1
Management Systems

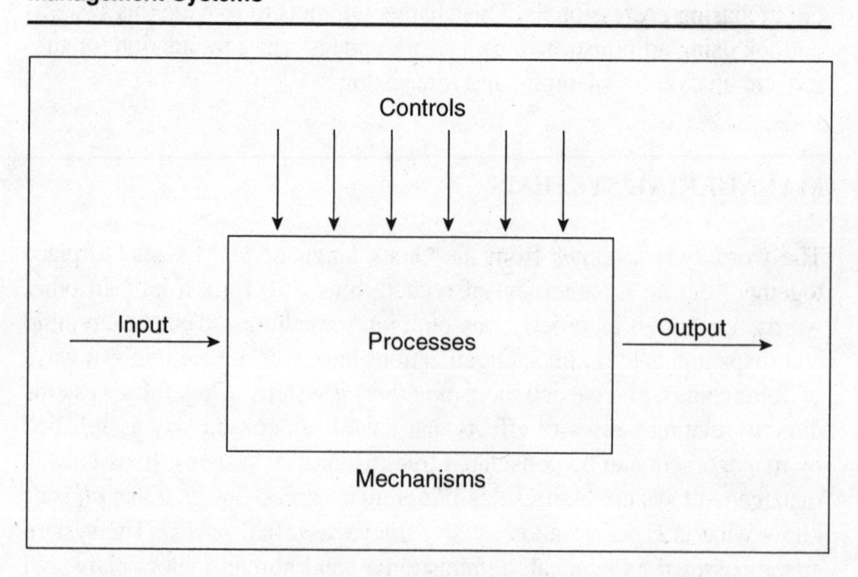

is repeated. A simplified depiction of a system showing how processes or mechanisms modify inputs into outputs is shown in Figure 3-1.

Reading modern writings in the field of management, it is possible to get the idea that the systems approach to management is something new. But it is not new. In fact, Frederick Taylor, the American father of scientific management, and Henri Fayol, the French father of administration management, advocated managerial systems at the turn of this century. "Systems thinking is a conceptual framework, a body of knowledge and tools that has been developed over the past fifty years, to make the full patterns clearer, and to help us see how to change them effectively."[1]

It is true, however, that computers and new mathematical techniques developed during the past two decades have made it possible to extend systems beyond the old limits and consequently to adopt some new systems. Notwithstanding these advances, the classic model of the management system is fairly consistent and includes philosophies and policies, principles and strategies, current practices, and records (described in Table 3.1). In this

TABLE 3.1
Classic Management Model

MANAGEMENT THEMES	QUESTIONS ADDRESSED	FULFILLMENT TECHNIQUES
Philosophies & policies	Why	Manuals
Principles & strategies	What, where, when & who	Departmental procedures
Current practices	How	Operating procedures, work instructions & individual practices
Records	Proof or documentation	Forms

[1]Senge, Peter M., *The Fifth Discipline: The Art & Practice of the Learning Organization*, New York: Doubleday, 1990, p. 7.

context, procedures are implemented to control the underlying processes, whereas records are used to show that the process in question is under control. In addition, forms typically fulfill a dual role: the form design is the document; when it is completed it becomes the record.

Managerial systems are a way of planning, organizing, directing, controlling, and evaluating work and workers, but they often are less effective at motivating workers. A good system becomes a plan for performing a task or a series of tasks. The Thomson Contracting System described in Chapter 2 is good example of a system put into place to ensure a task or series of tasks (i.e., annualized purchases). In addition, the system makes available in advance the resources needed to carry out the plan (organizing). The system becomes a way of directing the workers by describing the activities that are to be accomplished; similarly, it is a way of controlling the workers, since it can be used to compare what the workers are doing with what they should be doing.

Managerial systems are not usually motivational unless they provide real incentive. Unfortunately, managerial systems too often not only fail to motivate the worker, but actually reduce motivation. Diminished motivation arises when the system is unnatural to the worker (i.e., it is not the way the worker would accomplish the prescribed task if he or she were designing the system). Another discouraging aspect about systems is that they require a task to be performed in the same way and often at the same speed each time. This repetition is, of course, monotonous. A frequent complaint from workers around the world is that their work is boring.

In establishing an effective managerial system, one should try to achieve the following general objectives:

1. ***Increased profit.*** The system should make a contribution to profit. While the system will often reduce the cost of carrying out the function performed under the system, cost reduction is not always necessary to make a system financially advantageous. For example, a new system for handling customer complaints may actually cost more than the system it replaces, but it may add proportionately more revenue by contributing to customer satisfaction, which in turn generates additional sales.
2. ***Uniformity.*** A standard system makes it easier for users to understand and react. To illustrate, if the production planning department of a company published the production schedule on a different day each week, using a variety of formats, it would

be very difficult for the production manager to understand and use the schedule. However, if the planning department issued the production schedule on the first Monday of the month, two months in advance, and always in the same format, the manager's comprehension would be greatly improved.

3. ***Fixed responsibility.*** A good managerial system fixes responsibility for the functioning of the entire system.

In the planning stage, it is important to take into account that a good system:

1. Increases profit, is standardized, and assigns fixed responsibility.
2. Is flexible.
3. Feeds back needed information. A system that does not feed information back to management is called an open-loop system. A closed-loop system feeds information back as necessary. A good system does not necessarily have to be a closed-loop information system; however, if there is some important information that could be gained from the operations, then a closed-loop design would be better.
4. Provides internal control (when money, goods, or property are involved).
5. Facilitates the training of employees.
6. Facilitates automation, thereby reducing cost.
7. Is simple and easy to understand.

METHODS OF ANALYZING PURCHASING SYSTEMS

Managers responsible for managing purchasing administration and paperwork in an organization, whether they are a purchasing professional, retail buyer, office manager, controller, receiving supervisor, or store manager, must recognize what systems are necessary, know the systems well, see that they are followed, and try to improve their efficiency. In addition, the administration must cater to the requirements of internal customers and suppliers as well as satisfy the overriding needs of external members of the supply team. It is important to recognize the distinction between external and internal members:

> External customers purchase the product, financially supporting the organization. Obviously it is important to satisfy these people. Inside the company, employees pass on their work to other employees, who are their internal customers. Similarly, external suppliers are the people outside the organization who sell materials, information, or services to the organization. Inside the company, employees receive work passed on from other people in the organization, the internal suppliers.[2]

Each purchasing employee, therefore, is a customer of preceding workers and each is a supplier to succeeding workers in the chain.

In analyzing purchasing systems the following steps should be addressed:

1. The first step in analyzing a purchasing system is to examine the required information flows that are necessary to accomplish the purchasing organization's mission. Often, this analysis will focus on the interfaces between purchasing and internal customers within the organization as well as with members of the supply chain.
2. Using these detailed information flows, it should be possible to ask why they should be systematic and to develop a list of systems that the department should have. If unaware of what they should be, it may be necessary to consult a textbook or handbook.[3] Throughout this process, it may be useful to consult the controller, other purchasing professionals, key contacts in other departments and even sales representatives from companies that sell office forms. To gain objective and unbiased insight it may be necessary to discuss information flows with academics, consultants, or members of trade associations.
3. After making a list of needed systems, determine which systems are in place and which ones may be missing. It is highly likely that the systems presently in effect are the most important ones. They exist because of necessity.

---

[2]Scholtes, Peter R. *The Team Handbook: How to Use Teams to Improve Quality,* Madison, WI: Joiner Associates, 1988, pp. 2-5.

[3]Examples of such reference materials include *Purchasing and Materials Management,* Tenth Edition, by Michiel R. Leenders and Harold E. Fearon, Homewood, IL: Irwin, 1993 and *The Purchasing Handbook,* Fifth Edition, by Harold E. Fearon, Donald W. Dobler and Kenneth H. Killen, New York: McGraw-Hill, 1993.

4. The next step is to analyze the most important system that already exists. For most purchasing organizations, the purchase order–receiver system is often examined first.
5. In analyzing any system, the first task is to depict what is presently being done. There are three possible ways to characterize a paperwork system: (a) make a flowchart of the system, (b) make a diagram of the system, or (c) list the steps in the system.
6. After thoroughly understanding the existing system, it is necessary to compare it against an optimal system which is based on information flows. In conducting this gap analysis, it is necessary to determine if there are any steps in the current system that can be eliminated or combined.
7. Based on this gap analysis, the new purchasing system is developed. Additionally, one should confirm or verify that the proposed purchasing system contributes to making the supply chain work using previously-mentioned criteria.
8. After verification, the system should be implemented and validated under dynamic operating conditions prior to repeating the entire process.

Table 3.2 lists a very simple purchase order system and Figure 3-2 shows the same system in diagram form. Figure 3-3 shows the current receiving system. After analyzing the existing system by examining requisite processing flows and the information required, recommendations for

TABLE 3.2
Current Purchase Order System

	CURRENT PURCHASE ORDER SYSTEM
1.	The requesting department telephones order to purchasing.
2.	Purchasing types order.
3.	Purchasing mails Copy #1.
4.	Purchasing files Copy #2 in its follow-up file.
5.	Purchasing sends Copy #3 to the requesting department.
6.	The supplier acknowledges the order.

FIGURE 3-2
Current Purchase Order System

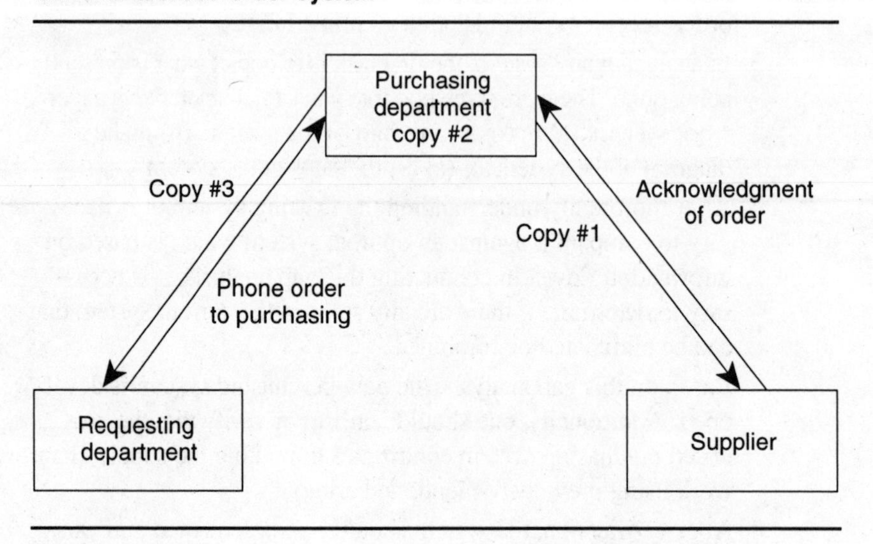

improving the system can be developed (Figures 3-4 and 3-5 show how these systems have been changed).

After visualizing the current system (often using diagrams, flowcharts or process maps), it is necessary to determine if there are any steps that can be eliminated or combined. An additional question to ask is, "Do we need to add anything to provide information to departments that need to know?" Computer screens will replace paper soon, but one will still need to understand how the system works.

In the purchase order–receiver example, the arrival of a receiver packet in the receiving department accomplishes two things. First, it notifies the department that the shipment is due, when it is due, by what mode of transportation, and where it is to be delivered once it has been received. Second, information accompanying the packet can help reduce duplication. The receiving department needs to know that the shipment is due for the following reasons:

1. ***To ensure that only the materials ordered by the company, division, or store are received.*** It is not uncommon for freight bills to have incomplete or wrong addresses, or, in the case of companies with multiple outlets in the same city, of having a shipment delivered to the wrong store. The wrong delivery creates confusion both before and after the error is discovered. A store that receives an

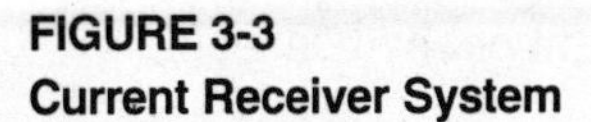

FIGURE 3-3
Current Receiver System

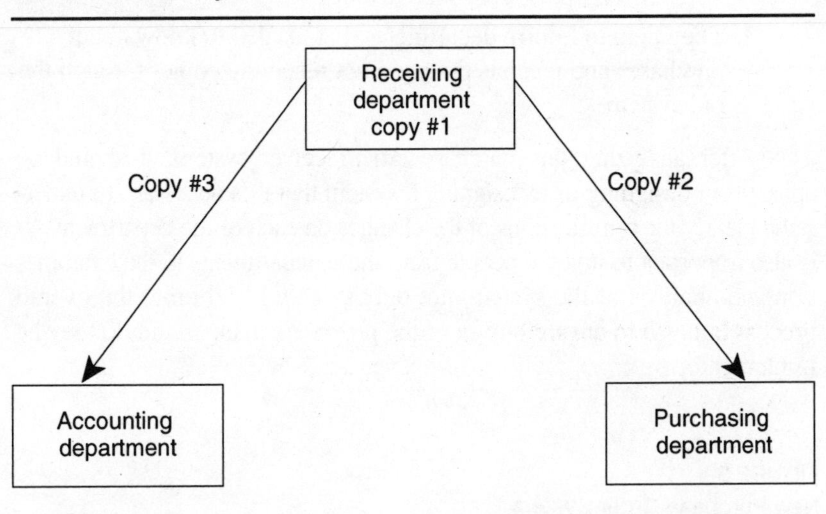

improper delivery does not want it and probably does not know what to do with it. All it can do is call the other stores and hope that it can locate the store that ordered the materials.In the meantime, sales are lost because the order was not delivered to the right location in the first place. Extra cost is also involved in locating the rightful owner, in changing the paperwork, and in the extra handling and shipping involved in transferring the shipment to the right location.

2. ***To provide information for planning purposes.*** By knowing what is expected and when, the receiving department can plan the department's work better.
3. ***To provide other information.*** The receiving department also needs to know where to deliver the incoming shipment and to what department the receiving costs should be charged. Since this information is (or should be) contained on the requisition, it can easily be typed on the purchase order/receiver, thereby making it readily available for receiving's use.

Summary of New Purchase Order/Receiver System

Figures 3-4 and 3-5 depict the new systems. These new purchase order/receiver systems are more effective for the following reasons:

1. They reduce cost for the receiving department by being able to schedule and plan work more evenly.
2. They help to inform departments that needed to know about purchases and receipts but were not receiving copies through the old system.

After analyzing this purchase order/receiver system, it should be apparent in designing or redesigning a system that it is necessary to understand clearly the ramifications of the changes on each of the departments. It is also important to involve people from these departments in the examination and analysis of the system, not only to gain insight into the overall process but also to ensure buy-in of the proposed changes should they be implemented.

FIGURE 3-4
New Purchase Order System

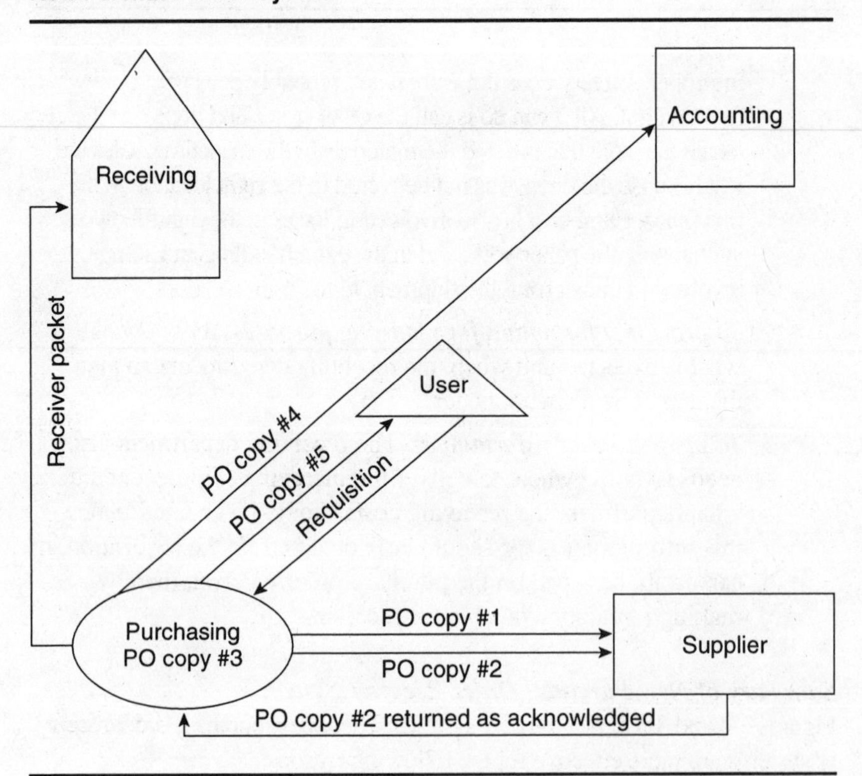

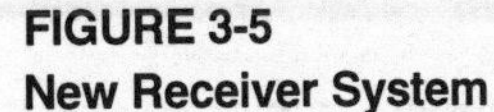

FIGURE 3-5
New Receiver System

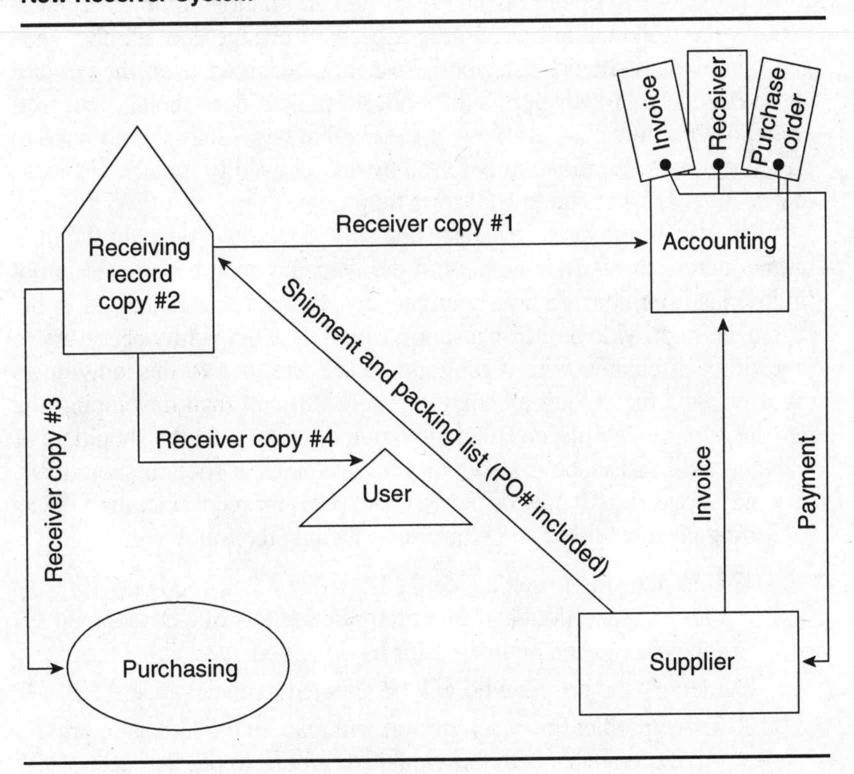

GETTING ACCEPTANCE OF THE SYSTEM

Whether proposing a new system or some changes in the old system, it is necessary to sell the ideas to other people in the organization, and oftentimes to someone outside the organization, such as customers or suppliers, who will be affected by the changes. Frequently, such changes are met with resistance. Reasons for resistance to change can be divided into three main causes:

1. *Economic;* such as fearing loss of benefits or job
2. *Personal;* do not want to make the effort to change or think there is an implied criticism of something they have been doing
3. *Social;* do not want to change locations or work with new people.

Need for New Ideas to Be Tested

Most resistance to change is based on human relations problems, rather than on the technical and economic aspects of change. Not all change is good; new ideas should be tested before they become part of the policies and procedures. However, this does not mean that ideas should be rejected simply because they are new. It means that new ideas should have to pass a rigorous examination before they are allowed to replace old ideas that seem to have operated well for a long time.

To be an innovative manager one should take into account that it is human nature to resist change. And the tendency to resist change is not limited just to unionized hourly employees. Often, other managers in the organization provide the strongest opposition. It is frequently necessary to precede each change with a program to sell the idea to the individuals involved, and the selling job may be more difficult than developing the new idea in the first place. True innovation often rests on the shoulders of a champion. Persistence is also required. If an idea is good, it should survive initial rejection. It may be necessary to press for reconsideration. Ideas for selling change within an organization include the following:

1. Before trying to sell an idea, ensure that it is a good one. By checking and double checking the soundness of a concept, an innovator gets a reputation for having good, workable ideas.
2. Identify the people who will be affected by the change.
3. Try to predict how each person will react to the idea. Use previous experiences with individuals or groups to predict reactions.
4. Using your predictions, work out an appropriate approach to each individual or group.
5. Approach the people (on an individual basis) who have the power to reject your proposed change, and feel them out on the subject. Try to find out how they stand on the issue without getting them to commit themselves. The purpose of this first meeting is to determine what it will take to sell each person who has a stake in the change. Everyone may not have to be sold, but chances are that a good many will.
6. In addition to selling those who have the power to veto proposed changes, it is necessary to convince other individuals who are affected by your idea. While they may not be able to overtly resist proposed change, their continued support may dictate whether the idea is truly implemented over the long term.

Sometimes a rational approach to selling is successful, while at other times an emotional approach, or a combination of emotional and rational approaches, works. Another method to getting a change approved is "horse trading." Offering to exchange something of importance to the other individual may make it possible to gain greater acceptance for an innovative idea. In some cases, it may not be possible to sell the concept. Perfectly good ideas do get rejected. It may be necessary to accept these defeats gracefully without becoming bitter or discouraged.

The biggest enemy to selling an idea is speed. Often before a concept is fully developed or the sales plan is articulated, an individual rushes breathlessly into the boss's office and announces this partly formed idea, only to have the boss shoot it down. Resist the temptation to be in a hurry to present new suggestions. Wait. Take time to develop and sell the idea. Another technique may be to convince someone else in the organization that it is his or her concept. In the long run, the only ideas that are going to be adopted are those that are put into practice and work well.

RECORD MANAGEMENT

A number of sources mandate a requirement for records. Such requirements should arise out of real need for information, but, all too frequently, many people spend time collecting information that is really not worth the effort. One of the main goals in making the supply chain work should be the prevention and elimination of efforts wasted on unnecessary record keeping.

The primary justification for record maintenance should arise from customers and suppliers. A secondary justification is that it offers information the purchasing department needs to operate well. Thirdly, consider the needs of the rest of the organization for information from purchasing to help the organization accomplish its goals. A fourth basis for record keeping is to satisfy internal and external auditors (i.e., records to be maintained as a "paper trail"). A fifth explanation for records relates to government regulators and taxers. A sixth basis is to be able to provide information in legal defense such as EEOC or liability suits.

TYPICAL PURCHASING DEPARTMENT RECORDS

Catalog File

This file should contain industrial directories, supplier catalogs, and sales promotion materials. Many organizations place a bookcase in the reception room and ask salespeople to keep their own catalogs up to date. This technique works especially well in small companies.

Open Files

The open file is where all records are kept about things that are pending or work-in-progress (WIP). This includes requisitions, quotes, firm bids, and purchase orders. Once an order has been satisfactorily received, the documents are destroyed or moved to longer-term storage.

Supplier Records

This is a file that contains basic information, such as the name, address, and phone number of key employees, history of their past performance, information about their union contracts, and any other information that would be helpful in future negotiations. It should also contain a record of meetings with suppliers (such as the date) and anything of importance that is discussed or decided.

Commodity Records

A record should be kept on each important material, product, or service purchased. A list of approved suppliers for each of these commodities should be part of this file. In addition, a description of the item or service involving references to drawings and specifications where appropriate should be part of this file. Some organizations keep information about suppliers and bid data in the commodity file, whereas others prefer to have separate files for suppliers and bid information. In any case, cross-reference files should be maintained. "When a firm uses a traveling requisition system for repetitively purchased inventory items, the traveling requisition and inventory record duplicate the commodity record. In such cases, there is no need to maintain a separate commodity record."[4]

In this context of record keeping, administrative tasks must be oriented toward adding value to the organization rather than being made up of nonvalue-added clerical activities. This chapter attempted to establish a systems outlook as it applies to purchasing administration using administrative paperwork and tasks as a foundation for the ensuing analyses of the planning and forecasting tasks.

---

[4]Dobler, Donald W., David N. Burt, and Lamar Lee, Jr., *Purchasing and Materials Management Text and Cases,* Fifth Edition, McGraw-Hill Publishing Company, New York, 1990, p. 66.

KEY POINTS

1. It is necessary to transcend the traditional reliance on existing purchasing administration to make the supply chain work effectively. Adopting a systems perspective that considers administration as it affects the entire supply chain rather than the purchasing department alone is an important step in competent supply chain management.
2. A system is an orderly way of doing something and can be anything that maps inputs to outputs. Only those systems directly relating causes to effects that have been consciously established by management can be considered true managerial systems.
3. In essence, a management system assures that *planning* is carried out in such a way that *all* staff know *what* is *expected* and *how* to achieve specified *results.*
4. Purchasing systems that allow an organization to receive requests to place an order, select suppliers, place the orders, track the orders, and receive the orders enable purchasing to support the firm's long-term supply needs and its production or service delivery systems.
5. The classic model of the management system includes philosophies and policies, principles and strategies, current practices and records.
6. Procedures are implemented to control the underlying processes, whereas records are used to show that the process in question is under control.
7. Forms typically fulfill a dual role: the form design is the document; when it is completed it becomes the record.
8. An effective managerial system increases profit, is uniform yet flexible, assigns fixed responsibility, feeds back necessary information, provides internal control, facilitates the training of employees, aids in automation, and is simple and easy to understand.
9. In analyzing purchasing systems, the process begins by examining the necessary information flows and interfaces that are needed to determine what systems are required. Subsequently, a prioritized gap analysis is conducted, comparing what exists to what is needed, and new systems are developed, verified, and validated.
10. Whether proposing a new system or changes in the old system, it is necessary to sell the ideas to other people who will be affected by

the changes. To be an innovative manager one should champion the idea but take into account that it is human nature to resist change.

11. The justification for record maintenance should arise from customers and suppliers, purchasing and organizational information needs, internal/external audit, and government regulator (or taxers), legal, EEOC, or liability requirements.
12. Typical purchasing department records include catalog files, open files, supplier records, commodity records, materials return forms, bills of material, tool and die records, and MRP schedules, as well as forms used for bid analysis, requests for quotation, inquiries, price records, supplier ratings, tracer requests, surplus equipment, and quality performance records.

REFERENCES AND RECOMMENDED READINGS

Leenders, Michiel R., Harold E. Fearon and Wilbur B. England *Purchasing and Materials Management,* Tenth Edition, Homewood, IL: Irwin, 1993.

Senge, Peter M., *The Fifth Discipline: The Art & Practice of the Learning Organization,* New York: Doubleday, 1990.

CHAPTER 4

PLANNING AND FORECASTING

CURRENT PURCHASING PLANNING

In making the supply team work, planning is the first and most important step because it lays the foundation for all subsequent management activities. The typical purchasing department has been provided a significant opportunity to leverage the supply chain for a competitive advantage in planning, primarily because it offers a chance to break with tradition. Profit and nonprofit organizations alike historically have not expected nor asked for proactive purchasing planning, and, instead, they have relied on purchasing to fill requisitions. Because of decisions (e.g., make-or-buy or product specification) by others in the organization, purchasing may be given insufficient time to order requisitioned materials. Such reactive requisition-filling situations necessarily limit the available options and lead to a firefighting modus operandi.

Management should expect purchasing to develop near-term, intermediate, and long-range operational and strategic plans. Top management must give purchasing the necessary authority and responsibility for planning.

How to Translate Theory into Practice—An Overview

How does one apply the theory of planning to the purchasing function? Quite simply, planning can be defined as a process of deciding in advance what is to be done, who is to do it, how it is to be done, when it is to be done, and how well it is to be done. As was stated earlier, purchasing planning should entail up to 50 percent of a purchasing professional's time, depending upon where he or she resides in the organization. The higher one's position in the hierarchy, the more time one should spend in planning.

Operational and Strategic Plans

Operational purchasing plans assure the continual flow of goods and services required for the day-to-day operations of the organization. Strategic purchasing plans support the goals and strategies of the organization and contribute to the development of new organizational goals and strategies. Both operational and strategic plans include (1) goals and objectives, (2) programs, (3) standards, (4) policies, (5) procedures, and (6) budgets. In this chapter we describe goals and objectives, programs, standards, policies, and procedures. Budgeting is addressed in the next chapter.

DEVELOPING PURCHASING OBJECTIVES

Organizational Values

Many organizations have found over the past decades that a statement of organizational values to guide all activities of the employees is useful. For example, the statement, "We believe that our customers, employees, and suppliers all need to be treated equally," is a valuable mandate, elevating the status of suppliers in organizational perception. Some organizations identify a corporate mission, others include organizational values as part of their objectives. For example, Hewlett-Packard's statement of corporate objectives, with regard to its own employees, includes:

> *Our People:* To help Hewlett-Packard people share in the company's success, which they make possible; to provide job security based on their performance; to recognize their individual achievements; and to ensure the personal satisfaction that comes from a sense of accomplishment in their work.

Purchasing objectives need to be congruent with organizational values and goals. Purchasing is in a unique position to ensure that value is contributed in the goods and services required by the organization. By brokering the information available to the purchasing department, buyers can dramatically enhance the success of the organization by getting the best value from its purchases. To capitalize on this opportunity, purchasing must have a well-articulated and readily executable set of plans. The planning process starts by establishing purchasing department objectives that support the organization's goals.

Setting Objectives

Objectives are general statements about expected outcomes. Organizational or departmental codes or mission statements are goals. Motorola's six sigma emphasis is an example of an organizational goal with respect to quality. The objectives of purchasing are to buy materials and services of the right quality, in the right quantity, at the right price, at the right time, from the right source. The primary operational objectives should include:

1. Buying at the lowest price, consistent with required quality and service.
2. Optimizing inventory turnover, thereby diminishing excess storage, carrying costs, and inventory losses as a result of deterioration, obsolescence, and pilferage.
3. Maintaining continuity of supply, preventing interruption of the flow of materials and services to users.
4. Maintaining the specified material quality level and a consistency of quality which permits efficient and effective operation.
5. Developing reliable alternative sources of supply to promote a competitive atmosphere in performance and pricing.
6. Minimizing the cost of acquisition by improving the efficiency of operations and procedures.
7. Hiring, developing, motivating, and training personnel and providing a reservoir of executive talent.
8. Developing and maintaining good supplier relationships in order to create a supplier (attitude and) desire to furnish the organization with new ideas and products and better prices and service.
9. Achieving a high degree of cooperation and coordination with user departments.
10. Maintaining good records and controls which provide an audit trail conforming to corporate audit objectives while assuring efficiency and integrity.

There are also secondary objectives which are supportive of other functions in the organization. These secondary objectives could be as follows:

1. Searching for effective new products that are relevant to the organization's operations and which can be called to the attention of users.

2. Suggesting to users materials or components that may improve the organization's products or operations.
3. Striving to introduce standardization in requirements to simplify specifications and reduce the costs of materials and types of inventory.
4. Furnishing data for forecasting or to assist in forecasting the availability of materials and trends in prices.
5. Adding value in make-or-buy decisions.
6. Serving sales personnel by bringing to their attention effective methods and techniques used by suppliers' sales forces and warning against undesirable conduct observed on the part of such personnel.

This list is not exhaustive, and each organization should incorporate specific purchasing objectives, tailoring them to their individual organizational needs.

Strategic Objectives of the Purchasing Department

The strategic objective of purchasing is to develop and monitor supply sources and supply relationships for goods and services in support of the organization's overall strategic goals. An effective strategic goal provides the purchasing organization with a mission that it can further define for its own purposes and that it can use to help suppliers and internal customers leverage the supply team's direct and indirect contributions for a sustainable competitive advantage. The challenge is to reduce this overarching set of goals into the various elements (or plans, policies, and procedures) that can be readily adopted by buyers and staff and that will result in successfully bringing the requisite goods and services into the supply chain while minimizing the resources required.

Several examples may illustrate the application of strategic objectives. As part of an organization-wide drive to improve profitability, a consumer goods organization's purchasing department reduced the number of MRO suppliers from more than 700 to one. This enabled them to: (1) reduce supply and procurement staff by five people and (2) improve delivery performance. Systems contracting and EDI were used to almost eliminate internal stores. Overall savings were approximately 15 percent of MRO purchases, which amounted to about 7 percent of all purchase dollars. These savings accrued directly to the bottom line, resulting in about a 9 percent improvement in profits.

In another project in the same company, marketing was eager to reduce its new product introduction time. Careful analysis showed that new package development was a major time consumer, at more than 12 weeks. The development of a printer-partner with high technical and implementation skills reduced the package development time to two weeks, giving marketing an important edge. It was estimated that this change alone allowed the company to increase the market share on new product introductions by close to 20 percent, which improved profitability to more than 40 percent!

How to Set Objectives

Objectives are targets set by management to give the organization direction. While setting objectives is a very old concept, emphasis on their use to drive business practices increased significantly after 1954 when Peter Drucker, in *The Practice of Management,* stressed the use of objectives as devices for planning and controlling. Drucker's book caused managers (particularly top managers) throughout North America to start managing by objectives. Many organizations still use management by objectives (MBO) programs that were initiated in the mid-fifties and early sixties.

Organizations that implement MBO effectively have each manager, from the first line supervisor to the president, establish written objectives. Each manager writes personal objectives, usually completed annually, and submits them to his or her superior for approval. In many cases, this process may be iterative and often involves compromise; but no matter how long it takes, it is essential that the final objectives be mutually agreed upon. After the objectives have been approved, they are normally consolidated, thus forming a set of objectives for the supervisor, and are sent up the chain of command. This process is repeated until the objectives reach the highest level decision maker, at which point they become the organization's objectives. During the year, usually quarterly and semiannually, each manager's superior reviews the objectives to ensure that the manager is making adequate progress.

Types of Objectives

George S. Odiorne has classified objectives into four types: (1) innovative; (2) problem-solving; (3) regular job duties; and (4) self-improvement.

The *innovative objective* is creative and results in novel ways to reduce costs or better ways to do things, or in new products or new markets. The *problem-solving objectives* set goals that will solve a problem. For example, too often office supplies are delivered late. A problem-solving objective resolving this late delivery situation might read, "within 60 days, office supplies will always be delivered on Tuesday between 9 A.M. and noon." *Regular job duties objectives* are written to clearly define the job results that will be used to measure a manager's performance. An example of such an objective is, "I will meet my department's budget each month." The *self-development objective* is aimed at getting managers to realize their weak points and make a plan to overcome them. "I will spend five hours a week reading trade and professional journals," might be an objective for a purchasing manager.

How To Write Objectives

In writing objectives, one must first decide what type of objective is needed. It is important to be specific, using quantitative statements whenever possible. Do not write, "I will do a good job on my budget." The trouble with this objective is that the phrase "good job" is not clear. Do write, "I will meet my budget," or "I will keep actual expenses 2 percent below my budget this year." Finally, set a definite time for completion. If the objective is complicated, with a number of parts, establish a deadline for each part.

Most organizations have their own forms to be used for writing objectives.

Fundamental MBO Concepts

Priority of objectives. Not all objectives should receive the same priority, so it is necessary to first pursue those that have the greatest impact on the operation of the organization. These priorities should be established and communicated throughout the organization to test whether or not the objectives and the assigned priorities are valid.

Interpretation of objectives. Purchasing objectives must work in conjunction with those of the rest of the organization. Integration of objectives assures success of the overall goal. It is imperative that all supply personnel understand the supply function's relationship to the organization it serves. Typically, the greater the value of the inputs as a percentage of total revenue, the greater will be the significance of accomplishment of supply objectives.

Measurability of objectives. Objectives also need to be divided into groups (e.g., cost savings, product improvement) that allow them to be quantified and measured. This is generally done by quantifying the existing situation, and then comparing the situation at some point in the future against historical data or appropriate external benchmarks.

Basic Steps in the MBO Process

The basic steps involved in conducting an MBO program as the basis for performance planning and evaluation activities are:

1. **Definition of responsibilities and preparation of the job description.** Each individual reviews, modifies, and updates his/her own job description. After the initial effort, this procedure typically involves "give and take" discussions with the supervisor, until a mutually acceptable decision is achieved.
2. **Establishment of individual objectives.** After the individual and the supervisor agree on the details of the job description, the individual is asked to plan how he/she will accomplish the results necessary to fulfill the job responsibilities. This requires the establishment of specific objectives within a designated time schedule. The planning period typically runs for six months or for one year.
3. **Agreement on objectives.** After preparation of the objectives has been accomplished in preliminary form, the individual discusses the tentative plan with the supervisor. During the discussions, the supervisor's role is that of questioner, developer, and counselor. During the entire process, it is important for the individual to feel ownership of the resulting objectives and at the same time there should be complete agreement with the supervisor concerning the objectives established.
4. **Establishment of evaluation criteria.** It is important that the individual and the supervisor jointly determine in precise, quantifiable terms exactly what checkpoints and criteria will be used in evaluating progress toward achievement of the objectives. Typical criteria include project due dates and formats, cost/profit figures, comparison of performance with historical trends or with other performance levels within the organization, etc.

5. **Comparison of performance with the plan.** The individual and his/her supervisor jointly review the performance and compare it with the plan (or standard). Performance must be evaluated as objectively as possible in light of objectives and expectations established in the prior planning process. From the results of this planning and evaluation process, subsequent plans can be developed to move toward further achievements in terms of both departmental and individual progress.

PROGRAMS OR PROJECTS

Programs or projects provide the means to achieve objectives and are particularly valuable where change from past practice is required. They involve planning for future events and establishing a sequence of required actions. They might include such general activities as locating new suppliers as a source of additional raw material, or auditing supplier quality, supply chain management, or just-in-time.

Programs or projects set forth a set of steps and a time table aimed at completing a specific task. Examples include the following: (1) Chrysler, Ford, and General Motors embarked on programs to reduce dramatically their supplier bases (and so have thousands of others); (2) Motorola, to name one of many companies, had a program to develop six sigma quality; (3) many other organizations have had programs or projects to set up a matrix structure, establish *Kanban* systems, or use the JIT approach to manufacturing and supply; and (4) currently many organizations are involved in EDI implementation and ISO 9000 registration, both of which are described below.

Electronic Data Interchange (EDI)

Many world-class purchasing organizations use paperless systems based on EDI, optical character recognition (OCR), automatic funds transfer (AFT), bar coding, and similar technologies. Because of the significant impact that EDI has had on purchasing administration and in making the supply chain work, it will be described in detail.[1] EDI is essentially the

---

[1] Information in this section has been compiled with the assistance of the Consortium for Advanced Manufacturing-International (CAM-I) Quality Customer/Quality Supplier (QC/QS) Program.

electronic exchange of business transactions in standardized formats. It is usually intercompany but can also be intracompany. Transactions are often batch, fast batch, or on-line, and are intended to be machine processed (not free-form messaging). Typical uses for EDI are shown in Table 4.1.

TABLE 4.1
Typical EDI Use

TYPICAL EDI USE	
Pre-order information	Catalog request Planning schedules Design specifications
Order information	Order instructions Purchase orders Order status Order change Order acknowledgment
Materials management	Materials release Advanced shipping notice Change notices
Transportation	Freight bill Bill of lading Packing slip Location inquiry
Financial	Invoice Remittance advice Payment Credit/debit memos

The Advantage of EDI

Some of the most frequently cited immediate advantages for implementing EDI are that it eliminates manual reentry of data, improves customer service, leads to better inventory management, helps to control cash flow, and can generate working capital. By eliminating mail float and enabling orders to be processed faster (no manual intervention) and more accurately (no redundant data entry), EDI improves customer service. Better inventory management (based on more accurate sales order forms, reduced purchase order lead time, and reduced requirements for safety stocks) also leads to higher service levels. Reducing the required inventory, accelerating the sales cycle, and enhancing more timely invoice processing contribute to cash generation and increased working capital.

Implementing EDI

EDI adoption is a major undertaking and should not be pursued merely in response to a trading partner's insistence, or because competitors are implementing EDI, or because a company is desperate for a quick-fix panacea. Before espousing EDI, an organization should ensure that exchanging information electronically supports the overall organizational strategy (including relevant sponsorship at the appropriate levels). Also relevant are the cost and ramifications of EDI standard tools and techniques (including implementation, software maintenance, manpower, and participant training) and how to promote systems and applications integration. An additional consideration is how to drive the necessary organization and process changes. Each supplier implementation is different, and it is necessary to work with software firms to tailor software packages. Obstacles to EDI are shown in Table 4.2.

Adopted effectively, EDI can help reduce variability in work processes, accelerate information exchange between members of the supply chain, and increase operational efficiency by lowering overhead costs. Thomson Consumer Electronics has used electronic RFQs for the past two years, which has precluded the use of paper. The supplier now has the obligation for most of the data entry, and subsequent communication occurs by E-mail, significantly reducing the cycle time. EDI also often provides a mechanism to exchange information that could not be readily supported by a paper environment (i.e., product information, forecasting, and inventory status to trigger replenishment). Most important, however, its use frees purchasing professionals from some of the more mundane clerical tasks, thereby affording them greater opportunity for building stronger trading partner relationships

TABLE 4.2
Obstacles to EDI

OBSTACLES TO EDI
Difficulty of quantifying long-term benefits
Independent, micro-managed decisions
Lack of commitment to change & required resources to support change
Changes to existing systems & processes
Integration of new systems
Continuing to think in terms of paper transactions

and, ultimately, for enhancing competitive service and market differentiation. In the final analysis, EDI is available now, is necessary to remain competitive, and is evolving rapidly. Its adoption takes management commitment, strategic planning, and cooperation between internal and external partners.

It should be noted that programs or projects are intended to accomplish a desired end result such as locating new suppliers or reducing the number of existing suppliers. In other instances, programs and projects simply establish an approach to something, such as a supply chain management. But once a program has been established and is inculcated into normal operating practices, it is no longer a program but a new way of managing.

STANDARDS AND BENCHMARKS

ISO 9000

Standards are used to measure either the quality or quantity of work accomplished. They are usually based on knowledge of what has been done in the past, or on a more scientific method. The most widely accepted system of standards currently extant that has a direct effect on the future of purchasing is the "ISO 9000-9004 Series" of Quality Management and Assurance Standards. The International Organization for Standardization, headquartered in Geneva, Switzerland, is the governing body for the ISO 9000 standards, a series of internationally accepted standards by which a company can ensure its own quality management system. These five standards were prepared by Technical Committee ISO/TC 176 on Quality Assurance in the interest of harmonizing the large number of national and international standards in this field. One hundred thirteen countries participated in the approval process, and by 1994, 82 countries had adopted national standards patterned after the ISO documents.

Businesses or organizations can have an operation, division, product line, or their entire company registered under these standards. Registration assures customers of an "operating" quality management system. Registration is obtained through a third party audit performed by a registration body, which must be approved by an international certification organization. By 1994, 22 of these registration bodies were operating in the United States. Registrations are rigorous and can be costly. However, it is acceptable to advertise registration, although registration symbols may not be used on a product, and symbols must be used in a manner authorized by the registrar.

In general, the advantages of ISO 9000 are shown in Table 4.3. Regarding specific benefits, DuPont, which has aggressively pursued ISO 9000 registration for its sites worldwide, identifies the following results as being directly attributable to the registration process[2]:

- A European site with greater than a 10 percent cost reduction and greater than a 10 percent increase in production output
- A site cost reduction of $3 million
- Milling cost as a percent of sales reduced from 60 percent to 50 percent
- On-time delivery increase from 70 percent to 90 percent
- Final assembly yield improvement from 92 percent to 96 percent
- Justified complaints down 26 percent
- Corrective actions down 62 percent

---

[2]Rita Granville, DuPont Corporation, presented at the 1992 National Association of Purchasing Managers Annual Conference, San Francisco, California, May, 1992.

TABLE 4.3
ISO 9000 Advantages

ISO 9000 ADVANTAGES
Improved quality systems
Reduced variability
Improved quality in supplied products
Reduced redundancy in auditing
Increased communication among & within departments
Time for partnership development
Time for continuous improvement activities
Increased training
Competitive advantage in marketing (short-term advantage)

The following sections have been extracted from ANSI/ASQC Standard Q91-1987 (technically equivalent to ISO 9001, but incorporating customary American language usage and spelling) and apply to purchasing in organizations that are seeking registration under ISO 9000.[3] In this context, the term *supplier* is construed to mean the firm that is seeking conformance to the

[3]"Quality Systems--Model for Quality Assurance in Design/Development, Production, Installation, and Servicing," American National Standard (ANSI/ASQC Q91-1987), *Milwaukee: American Society for Quality Control,* 1987, pp. 3-4.

FIGURE 4-1
ISO 9000 Purchasing Interfaces

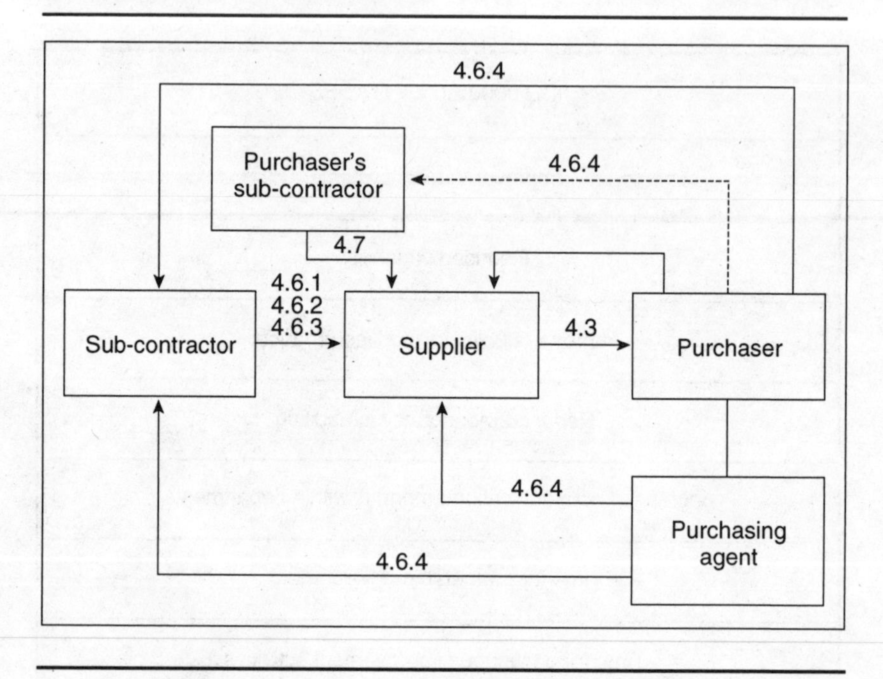

standard, while *sub-contractor* is used to describe members of the supply chain and *purchaser* denotes the external customer. Notwithstanding these directly applicable standards, purchasing will necessarily be involved throughout the ISO 9000 registration process. The interaction of these sections as they apply to purchasing administration is shown in Figure 4-1.

ANSI/ASQC Standard Q91-1987

4.6 Purchasing

4.6.1 General

The supplier shall ensure that purchased product conforms to specified requirements.

4.6.2 Assessment of Sub-Contractors

The supplier shall select sub-contractors on the basis of their ability to meet sub-contract requirements, including quality requirements. The supplier shall establish and maintain records of acceptable sub-contractors. (see 4.16 [Quality Records]).

The selection of sub-contractors, and the type and extent of control exercised by the supplier, shall be dependent upon the type of product and, where appropriate, on records of subcontractors' previously demonstrated capability and performance.

The supplier shall ensure that quality system controls are effective.

4.6.3 Purchasing Data

Purchasing documents shall contain data clearly describing the product ordered, including, where applicable:

a) the type, class, style, grade, or other precise identification;

b) the title or other positive identification, and applicable issue of specifications, drawings, process requirements, inspection instructions, and other relevant technical data, including requirements for approval or qualification of product, procedures, process equipment and personnel;

c) the title, number, and issue of the quality system Standard to be applied to the product.

The supplier shall review and approve purchasing documents for adequacy of specified requirements prior to release.

4.6.4 Verification of Purchased Product

Where specified in the contract, the purchaser or the purchaser's representative shall be afforded the right to verify at source or upon receipt that purchased product conforms to specified requirements. Verification by the purchaser shall not absolve the supplier of the responsibility to provide acceptable product nor shall it preclude subsequent rejection.

When the purchaser or the purchaser's representative elects to carry out verification at the sub-contractor's plant, such verification shall not be used by the supplier as evidence of effective control of quality by the sub-contractor.

4.7 Purchaser Supplied Product

The supplier shall establish and maintain procedures for verification, storage, and maintenance of purchaser supplied product provided for incorporation into the supplies. Any such product that is lost, damaged, or is otherwise unsuitable for use shall be recorded and reported to the purchaser (see 4.16 [Quality Records]).

> NOTE: Verification by the supplier does not absolve the purchaser of the responsibility to provide acceptable product.

General requirements imposed by section 4.6 of the standard are intended to ensure that sub-contractors (or suppliers) are able to deliver what is ordered. In this context, the standard calls for assurance that purchased products or services conform to specified requirements, assessment of sub-contractors, maintenance of purchasing data, and verification of purchased product. It is important to note, however, that the intent of the standard is to put the responsibility for compliance on the purchasing organization, ***not*** solely on the supplying firm (i.e., to ensure that items are described properly, requirements are accurate, or other interactions between members of the supply chain are coordinated effectively). According to the Victoria Group, which provides ISO 9000 consultancy services, pre-assessment audits, and in-house training[4]:

> There are a number of ways in which purchasing activity can be managed to eliminate these problems, and the accurate detailing of purchasing orders is one of them. Using the right sub-contractors is equally as important, hence the requirement for all sub-contractors to be assessed in some way before you use them.

Overall, the essence of ISO 9000 is to "Say what you do, do what you say, record what you did." The standards are generic and apply to any industry; they specify the necessary elements of a quality system (like a shopping list) but do not specify how to implement these standards. They do not deal with product certification, but instead serve as a quality system certification. Much of ISO is based on documentation, with third party (or independent) auditors determining if an organization is doing what they say they are doing to ensure that their quality system can be used for external quality assurance purposes. Based on past experiences with companies seeking ISO registration, the 85/15 rule appears to apply. If an organization makes the decision to pursue registration, it should expect 85 percent of the effort to be accomplished by the people doing the work and 15 percent to be done by the staff in coordinating the preparation, pre-assessment, and assessment activities. There is understandably a great emphasis on getting purchasing professionals involved in the ISO 9000 registration process from the outset.

---

[4]"Quality Assurance System Assessments and Lead Auditor Training Manual," Section 3, The Victoria Group, 1994, pp. 21-22.

Benchmarks

Currently, the term *benchmark* is being used by many to denote a standard. The Center for Advanced Purchasing Studies (CAPS) defines a benchmark as a standard or point of reference in measuring or judging quality or value. CAPS further states that benchmarks give purchasing professionals the reference point they need to evaluate their own firm's performance.[5] A difficult issue then is how to define standards or benchmarks for an organization. The place to start is with published benchmark studies on the industry. The CAPS benchmarking studies include many, but not all industries. An example from the *U.S. Banking Industry Study of 1993* is provided below.[6]

1. Functions for which purchasing is responsible:
   a. Purchasing (100 percent).
   b. Print shop (31 percent).
   c. Inventory control (77 percent).
   d. Disposal of fixed assets (62 percent).
   e. Warehousing (85 percent).
   f. Receiving (77 percent).
   g. Forms management (77 percent).
   h. Other (54 percent).
2. Total purchasing dollars as a percent of corporate asset dollars equals .29 percent (range from .07 percent to .98 percent)
3. Average dollars in inventory as a percent of total purchase dollars equal 2.3 percent (range from 0 percent to 17.2 percent)
4. Percent of total purchase dollars processed through EDI equals 5.1 percent (range from 0 percent to 20 percent)
5. Average purchase order cycle time for capital purchases equals 11.2 days (range from 2 to 45 days)

Benchmarks are often effective as static goals or objectives, whereas process benchmarking refers to the continuous search for and application of

---

[5]Stanley, Linda, Julie Gentry, Debra S. Seaman, and Rodney Doerr, *Purchasing Performance Benchmarks for the U.S. Aerospace/Defense Contracting Industry,* Center for Advanced Studies, Tempe, AZ, 1991.

[6]Campbell, Larry, Linda Grass, Charles Edmund Page, Dominion Bonksbores, and Paul C. Shaup, *Purchasing Performance Benchmarks for the U.S. Banking Industry,* Center for Advanced Purchasing Studies, Tempe, AZ, 1990.

significantly better practices that lead to superior competitive performance.[7] Addressing the issue of how an organization compares to leading-edge firms should lead an organization toward the development of industry best practices for its own internal operations.

Before beginning process benchmarking, an organization needs to take a "hard look" at several questions, including:

1. Are we willing to make a major change?
2. Is the expected improvement worth the expenditure?
3. Are the results important to us?
4. Does the process impact a critical success factor?
5. Have all investigations relating to this process been completed? Have all other alternatives been explored?
6. Have we begun measuring the current process?
7. Do we know the major cost components?
8. Are we willing to wait for a benchmarking study to be completed before implementing any changes?
9. Are we willing to reveal information about our own processes with outside organizations?

Process benchmarking may actually involve the division of the functions of an organization into several modules, analyzed as independent processes. Such a study may find, for example, that one organization has the best order processing, another has the best inventory control, and so on.[8]

Additional sources of benchmarking information include literature searches of industrial journals and purchasing magazines, internal studies to find best practices (usually only applicable if an organization is large with a number of purchasing departments), or industry studies. The latter studies are most effectively conducted through existing trade organizations or consortia such as CAPS or the Consortium for Advanced Manufacturing–International, or are undertaken by consulting firms or educational institutions.

---

[7]Leibfried, Kathleen H.J. and McNair, C.J., *Benchmarking: A Tool for Continuous Improvement,* New York: Harper Collins, 1992.

[8]For further reading on process benchmarking, please look at Hammer, Michael and James Champy, *Reengineering the Corporation: A Manifesto for Business Revolution,* New York: Harper Collins Publishers, Inc., 1993.

POLICIES

Policies are guidelines or general limits within which management acts. Policies exist at various levels of the organization—corporate, divisional, and departmental. They are valuable because they allow lower levels of management to handle problems without going to top management for a decision each time.

When the same decision comes up often, policies should be written down and distributed to all the managers involved. Or when management wants to head off a specific problem, policies are formulated to handle it "just in case." Examples of policies at various levels of the company might include:

1. No employee will be allowed to accept gifts from our suppliers.
2. Purchases for $100,000 or more must be approved by the CEO.
3. Purchases of $100 or less can be made from petty cash and do not need a requisition or purchase order.
4. Purchases of up to $500 can be made by a department head using a company-issued purchasing credit or debit card. No requisitions or purchase orders are needed.
5. All payment terms are to be net 60 days.

The main trouble with policies is that they tend to be inflexible. Employees too often follow policies that do not make sense under the circumstances. Good policies should encourage, not discourage, employees from using the chain of command for a special decision on a particular situation.

The growth of employee empowerment has meant that many organizations reduced the number of policies. Over the years the organizational rule book has grown (particularly in large private and public organizations) so that many policies are in conflict with one another. Refocusing on customer needs and allowing employees to use personal initiative (within broad guidelines) serves the customer best. Purchasers have to be alert that purchasing policies support organizational goals and strategies.

PROCEDURES

A procedure is a systematic way of handling regular events. It is stated in terms of steps to be followed in carrying out work and is used to control processes. Records provide evidence that the process described by the procedure is actually under control. For example, a bank teller has a procedure to follow when receiving funds. And an airplane pilot has a procedure for checking out a plane before taking off. In other words, a procedure is a list of systematic steps for handling events, situations, or decisions that occur regularly. Formal procedures have been committed to paper and have received organizational approval. Informal procedures have been developed for those situations in which no formal procedure exists.

The lack of procedures may harm an organization. An example is lack of proof that suppliers who are doing work on the buyer's premises have adequate liability insurance and appropriate worker's compensation coverage. Another procedural gap is to let people in shops or offices charge items at a local retailer without requisition approval. A procedure is required in both of these situations to protect the organization's assets. In most cases, exception-based purchasing controls are available as checks and balances for those attempting to bypass formal systems as well as to provide a rational and logical basis to ensure that supplier selections are based on least total cost-competitive, negotiation-based purchasing.

Special purchasing procedures and forms should be used for small value, usually nonproduction, supply-type goods and services, which often comprise the majority of POs but constitute a proportionately smaller amount of the dollar purchases.

FORECASTING

A forecast is a prediction of future events, and forecasting ***always*** relates to the future. It is different from planning in that it represents an attempt to predict future events, situations, or conditions. In the materials management context, two major relevant forecasts deal with requirements and markets.

Requirement Forecasts

Requirement forecasts are based on the organization's internal decisions and are essential to effective procurement planning. We need to know what we will need in the years ahead, how much, and when. Simple forecasts based on the pattern of demand for a product or service over time extend the past into the future and assume no changes in requirements other than in volume. For example, the bicycle manufacturing example provided later in this chapter falls into this category. More intricate time series forecasts take into account averages, trends, seasonal influences, cyclical movement, and random error.

More complex forecasts reflect factors affecting demand such as organizational goals and strategies and program requirements arising from them. For example, a hospital planning to expand its geriatrics services will require additional facilities, equipment, and medical supplies, all of which will require purchasing involvement.

Not all forecasts are accurate because no one can predict the future. It is often said that forecasts have only two outcomes: wrong and damn wrong. Events beyond the control of the forecaster may change plans and programs. For example, a lack of funding may delay the introduction of geriatric service or an earthquake may close the hospital. The economy of the nation is a common influencer of many organizational plans. Interest and inflation rates and the rate of growth of the economy tend to be watched very carefully by forecasters because of their impact on organizational plans.

Market Forecasts

Predicting market conditions into the future is the other critical dimension of forecasting for purchasers. Knowing what organizational requirements are likely to be is only part of the issue. Whether the market will be able to supply those requirements in the future is also important. And future market conditions (external factors over which management exercises little, if any, control) often indicate the direction of change in demand and frequently affect availability and price. For organizations with increasing requirements, potential future shortages are particularly challenging. For organizations depending on price stability, future increases or decreases in price may represent a significant risk. Purchasers of commodities are well aware of how supply and demand can affect availability and price, and they try to protect their organization by using appropriate strategies. Common factors affecting the demand for goods and services are shown in Table 4-4.

The *Annual Averaging Method* forecasts future item usage based on an average of past usage. It is used with the assumption that past usage is not subject to radical fluctuations and is a fairly reliable base from which future demands can be established. Just as the *Annual Averaging Method* assumes that future requirements will bear a certain relationship to the average requirements of past years, the *Simple Proportion Method* makes the same assumption about the relationship between past and future quarters

TABLE 4.4
Factors Affecting Demand for Goods and Services

EXTERNAL FACTORS	INTERNAL FACTORS
General state of the economy	Product or service design
Government actions	Price & advertising promotions
Consumer tastes	Packaging design
Public image of product	Salesperson quotas or incentives
Competitor actions	Expansion or contraction of geographical market target areas
Availability & cost of complementary products	Product mix
———	Backlog policy

Source: Lee J. Krajewski and Larry P. Ritzman, *Operations Management: Strategy and Analysis*, Third Edition, Reading, MA: Addison-Wesley, 1993, p. 433.

and their usage requirements. Specifically, the *Simple Proportion Method* assumes that the projected quarter requirements will have the same relationship to the first quarter of the year preceding it as the fourth quarter of the preceding year has to the fourth quarter of the year preceding it.

Trend Projection by Moving Average is a forecasting method useful when past usage has dramatic fluctuations, either seasonal or economic, which must be taken into account in order to estimate future usage more accurately. The *Prime Factor Method* forecasts requirements by comparing company needs to an outside standard. The amount of a product that an organization will use can be predicated by its correlation to industry use. For example, if an automobile manufacturer estimates increased sales for the next model year, then a company that makes parts for those automobiles can reasonably expect to have increased the output on its part also. The *Prime Factor Method* forecasts the demand for a given item a company can anticipate based upon industry expectations.

Qualitative considerations in the form of adjustments to forecasts are often made because of the influence of anticipated events and conditions that can affect projections. In particular, one should look at the analysis of forecasting errors, anticipated developments in other departments, business trends, and economic trends. This projection undergoes control when it is compared with actual demand and is subject to expert judgment based on internal and external factors before a modified prediction of the future needs is developed. Forecasts should be evaluated on the basis of losses or profits attributable to them.[9]

Regardless of what technique is employed, forecasts are predicated on the accuracy of the data. Historical data should represent the initial input and be plotted and examined for trends and seasonality. Outliers should be identified, and if a special cause for the outlier can be determined, it should be removed from interest. Eighty-five percent of the time dedicated to forecasting is spent gathering the data, and it is essential to establish a system to capture demand accurately (i.e., taking into consideration stock outs or limited supply).

Many forecasting software packages are available for microcomputers, and quantitative forecasting techniques have consistently been shown (using experiments with human subjects and ex post studies with company forecasts) to outperform judgment techniques. There is no clear winner between causal (regression) and time series approaches, although generally, time series methods are only applicable over the short range (less than 18 months). Finally, combining forecasts is usually more accurate than adopting individual techniques. Some generalized forecasting advice is provided in Table 4.5.

As mentioned in the first book in this series, purchasing strategy allows organizations to move from the present to the future.[10] The final book in this

---

[9]Purchasing professionals who are interested in more information about forecasting and forecasting techniques are advised to examine operations textbooks or manuals.

[10]Leenders, Michiel R., *Purchasing: The Acquisition Process,* Homewood, IL: Irwin Professional Publishing, 1994.

TABLE 4.5
Forecasting Truisms and Suggestions

Forecasting Truisms and Suggestions
Forecasts are *always* wrong
Forecasts of aggregate quantities are more accurate than forecasts of individual quantities
Forecasts are less accurate the further in the future they are
Forecasting is difficult (especially if it involves the future)
Forecasts should explicitly recognize their uncertainty
A forecast should not be confused with a plan
A single, consistent forecast should be developed
Demand should not be forecast when it can be derived

series covers the basic and most commonly used techniques of requirements and market forecasting. Forecasting future needs and future markets is essential to assure that the appropriate purchasing plans can be developed.

Materials Planning or Budgeting

The sales or revenue forecast is the basis for the materials forecast. An example involves the purchasing manager for a bicycle manufacturer. Last year, the company shipped about 1,000 units per month or 12,000 units. The economy is expected to grow about 3 percent this year and sales are anticipated to increase by 6 percent. Table 4.6 provides an example of a materials forecast.

SUMMARY

To administer purchasing effectively, purchasing management must develop near-term, intermediate, and long-range operational and strategic plans. Planning is essential to making the supply team work because it lays the foundation for all subsequent management activities. The strategic firm must include purchasing and supply considerations in its planning, development, and operations. It is also important to develop a purchasing managerial system that emphasizes planning and forecasting prior to developing budgets and financial strategies. Still, as developed in the next chapter, these considerations are often intricately linked and together lay the foundation for subsequent management options.

TABLE 4.6
Materials Forecast

	LAST YEAR	THIS YEAR	NEED PER MONTH
Handlebars	12,000	12,720	1,060
Wheels	24,000	25,440	2,120
Seats	12,000	12,720	1,060
Pedals	24,000	25,440	2,120
Frames	12,000	12,720	1,060

KEY POINTS

1. Planning is the most important step in the development of an effective purchasing department.
2. Operational and strategic purchasing plans both include goals and objectives, standards, policies, procedures, and budgets.
3. Operational plans provide for the daily needs of the organization, while strategic plans support the goals and strategies of the organization and help in the development of new goals and strategies.
4. Purchasing objectives need to be congruent with the organization's values and goals. These objectives must be tailored to meet the current and future needs of the organization.
5. Strategic objectives and goals provide a mission for purchasing to develop its internal and external resources into a sustainable competitive advantage for the organization.
6. Objectives must be clearly written, with easily measured goals and a timeline for completion. These objectives must be prioritized to best meet the needs of the organization.
7. An MBO program must define individual responsibilities and establish individual objectives. These objectives must be agreed upon by the individual's supervisor, and evaluation metrics must be determined. Performance of the plan should be compared with these metrics to determine the plan's progress.
8. Programs are a method through which objectives are achieved. They are a series of steps along an objective's timeline to completion.
9. EDI increases accuracy and reduces the delay time in the processing of purchasing orders. Increased information exchange between members of the supply chain reduces the time required for clerical duties (i.e., "paper pushing").
10. Standards based on previous knowledge and experience are used to measure the quality and completeness of a program.
11. ISO standards 9000-9004 apply to all members of the supply chain. ISO 9000 can be summarized as "Say what you do, do what you say, record what you did."
12. Benchmarks are used as a reference point or as a "yardstick" to evaluate performance of a plan. Relevant benchmarks for an

organization can be determined internally, or through external sources, including CAPS, CAM-I, consulting firms, or educational institutions.

13. Policies are guidelines that are laid out to reduce the need for higher managerial involvement in routine decision making.
14. Procedures are defined steps to be followed for resolving routine problems or performing regular operations.
15. Forecasts are based on internal decisions and historical performance to predict future needs. Factors including organizational goals and strategies, as well as external factors including inflation rates and economic growth, must be evaluated in the preparation of a forecast.
16. Predicting future market conditions is as important in forecasting as predicting internal demand. Product availability and price stability pose significant risks.
17. Regardless of the forecasting method used, accuracy of the predictions are directly related to the quality of the historical/environmental data.

REFERENCES AND RECOMMENDED READINGS

Burt, David K., Warren E. Norquist, and Jimmy Anklesaria, *Zero Base Pricing: Achieving World Class Competitiveness through Reduced All-In-Costs,* Chicago: Probus, 1990.

Drucker, Peter, *The Practice of Management,* Harper & Rowe, Publishers, New York, 1954.

Leenders, Michiel R., Harold E. Fearon, and Wilbur B. England, *Purchasing and Materials Management,* Tenth Edition, Homewood, IL: Irwin, 1993.

CHAPTER 5

BUDGETS AND CASH CONCEPTS

BUDGETING

The five basic functions of any business enterprise are: (1) producing a *good* or *service;* (2) marketing the good or service; (3) financing the operations; (4) purchasing goods and services from suppliers; and (5) staffing the organization. Even in a nonprofit organization such as a college or hospital, these activities are present. This chapter addresses the concept of financial planning as it applies to supply management.

Management typically uses a near-term financial assessment known as a business or financial plan (manufacturing concerns) or annual plan (non-profit organizations) to project income, costs, and profits. Budgets showing sources and allocations of funds usually accompany such plans, which essentially coordinate the financial requirements for operations, finance, marketing, and human resource and information strategies. Each plan or budget reflects time by identifying a planning horizon, usually from one to three years. These plans provide the overall framework for subsequent financial planning.

To finance an enterprise effectively it is necessary to know:

1. How much money is needed
2. When the money is needed
3. From where the money will come
4. Who will spend it
5. That the money is properly spent

A budget, then, is a financial plan that will help ensure that the five conditions listed above are met. In addition to being a financial plan, a budget should also provide a means for controlling expenses. First, the budget is in many cases the manager's authority to act. The manager can do anything that he or she has the budget to do. Conversely, the manager cannot spend money that has not been allocated (set aside) in the budget.

Second, the budget becomes a standard for measuring purchasing performance. Suppose that a purchasing manager's standard materials cost per unit (i.e., the amount budgeted for materials for each item produced) is 50 cents. If the manager spends 52 cents per unit, he or she is over the budget allowance. This manager could be doing a poor job on this budget item, which should signal the need to determine how materials cost could be brought down to the standard cost. Notwithstanding this analysis, the converse is not always true; there are some exceptions to the statement that a manager who is below budget is doing a good job while the manager who exceeds budget is doing a poor job.

Although this example describes the use of a budget for short-run control (monthly or quarterly), budgets are also used for long-term control. When budgets are developed, each account is typically compared with expenditures for the current year and previous years. When the new budget is adopted, the monthly report includes a column showing the cumulative amounts allocated or expended (actual year to date) compared with how much had been spent for the same account during the same time horizon for the previous year (last year's actual year to date). An example of such a budget is shown in Table 5.1, columns 7 and 8.

BUDGETS AND COST CONCEPTS

Although all or some of the following budgets may apply to other business functions, specific reference here is to their application to the purchasing department. There are five major types of budgets: (1) sales or revenue budgets; (2) materials or goods budgets; (3) expense budgets; (4) capital budgets; and (5) cash budgets.

1. ***Sales or revenue budget.*** Usually prepared by sales or marketing, the sales budget forecasts income based on expected sales during the coming year; in a nonprofit organization, this budget typically forecasts expected revenue. As mentioned in Chapter 4, the sales or revenue budget should be the first step in the overall budgeting process, because it aggregates overall requirements and provides the foundation for the other four budgets. In other words, the need for materials or goods (materials or goods budget), the amount to be spent for labor and overhead (expense

TABLE 5.1

Bank-In-A-Box Inc. Expense Reports for Dept. #10–Purchasing Department (Month Ending July 31)

1. Account number	*2.* Account name	*3.* July budget	*4.* July actual	*5.* July favorable (or unfavorable)	*6.* Budget year to date	*7.* Actual year to date	*8.* Last Year Actual Year to Date
10-1	Salaries	29,166	28,949	217	184,162	183,021	180,954
10-2	Fringe benefits	7,250	7,340	(54)	50,750	51,003	50,641
10-3	Non productive labor	500	597	(97)	3,500	4,021	3,236
10-4	Supplies	1,000	901	99	7,000	6,915	6,924
10-5	Maintenance and repair	2,400	1,836	564	16,800	14,779	17,177
10-6	Overhead allocation	3,000	2,975	25	21,000	26,749	20,557
10-7	Rent allocation	3,000	2,975	25	21,000	26,749	20,557
10-8	Light & heat allocation	200	149	51	2,700	2,414	2,501
10-9	Depreciation of equipment	7,000	7,000	0	49,000	49,000	49,000
10-10	Travel	3,000	3,102	(102)	21,000	21,100	20,500
10-11	Phone	2,200	2,196	(4)	14,400	14,379	16,500
10-12	Postage	100	100	0	703	700	700
10-13	Part-time help	1,500	1,694	(194)	10,500	11,200	10,012
	TOTALS	**61,316**	**59,778**	**530**	**403,515**	**414,030**	**398,759**

budget), the need for equipment (capital budget) and the anticipated cash flow (cash budget), are all dependent on the amount of the sales budget.

2. Materials or goods and supplies budget. This budget projects the amount and cost of materials (manufacturing or service operations) or goods (retail or wholesale firms) that need to be purchased during the budget period. The materials or goods budget is based on the sales or revenue budget. It is prepared by finding out from the sales or revenue budget *what, how much,* and *when* materials or goods will be needed on a monthly basis. To determine the annual budget, the cost of each item is multiplied by the amount needed per year. The primary purpose of a materials budget is to identify how much money will be needed to pay for purchases during the budget year. A properly prepared direct materials budget provides management with a tool that:

1. Establishes a basis from which the treasurer can estimate the financial requirements of the purchasing department
2. Requires the purchasing department to set up a purchasing schedule that assures delivery of materials when needed
3. Leads to the determination of minimum and maximum levels of raw materials and finished parts that must be on hand

Although it is generally based on estimated prices and planned schedules, the direct materials budget can:

1. Permit planned maximum lead time
2. Lead to an enhanced selection of sources
3. Reduce transportation costs
4. Provide a basis for planning workloads
5. Help in forward buying

In addition, the direct materials budget may improve purchase negotiations by reducing the pressure of time constraints. Frequently, the materials budget is compared to a standard cost. In this regard, purchased material variance is the difference between material value at standard cost and material value at actual cost. It involves two components: material price variance (MPV) and material quantity variance (MQV). MPV is equal to the volume purchased times the difference between actual purchase price and standard cost. The formula is as follows:

MPV = Volume * (Standard Cost - Actual Purchase Price)

MQV is equal to the standard cost times the difference between the standard quantity and the actual quantity. The formula is as follows:

MQV = Standard Cost * (Standard Quantity - Actual Quantity)

Since purchasing establishes standard costs and purchases materials, it is directly responsible for material price variance. However, purchasing is only indirectly responsible for material quantity variance, through such actions as the processing of poor quality items. MQV is generally the direct responsibility of the operations department. Purchased Material Variance (PMV) is a combination of MPV and MQV, whereby PMV = MPV + MQV.

*3. **MRO Supplies Budget.*** The organization also needs Maintenance, Repair, and Operating Supplies (MRO). Since the number of individual line items is likely to be so large that budgeting each item will not be feasible, such a budget is usually based on the use of past ratios and is adjusted according to anticipated changes in inventory and general price levels. Although purchasing should participate in the development of the MRO budget, and needs to know what quantities will be needed and approximately how much money will be spent, purchasing rarely incorporates these items into the purchasing department budget. Materials and supplies are budgeted in the department that will use or consume them. Still, purchasing must budget money for its own supply needs.

*4. **Capital budget.*** The capital budget specifies the amount of money to be spent for plant and equipment and can include anything from roof repairs to a new elevator, new cash registers, or hospital beds. This budget is usually allocated by having each manager submit a prioritized list of capital items and their estimated costs. The list allows finance to estimate how much cash will be required. However, at this point, no approval is given for any item. For each item listed, the manager responsible must justify the capital item targeted for purchase, which usually means demonstrating that this item is an absolute necessity or that it will return 20 percent or more annually on the investment.

The capital budget items may or may not have a direct relationship to the sales budget. Some capital items follow directly from the annual sales budget. When a new factory is built or a new production line is started, it is usually associated with an increase in sales or a new product line.

On the other hand, a capital item may not be a direct result of the sales budget. For example, replacing an old elevator in a department store may be necessary to meet government safety standards; or replacing a hand dolly with an electric lift truck in the storeroom may be desirable to reduce labor costs.

Capital expenditures are treated separately because of the magnitude of funds involved and the length of time required for capital recovery. Capital expenditures budgeting involves both short- and long-range expenditures. Short-range expenditures must be included in a budget for the current year and must be evaluated in terms of their economic worth. Long-range expenditures are not usually implemented during the current budget period; hence, their inclusion in the budget can be in somewhat general terms.

5. Expense budget. Sometimes called the operating budget, the expense budget is based on anticipated operating and administrative workloads, and it includes the expenses incurred in the operation of the purchasing function. Such expenses include salaries and wages, space costs, heat, electricity, telephone, postage, office equipment, office supplies, EDP costs, travel and entertainment, educational expenses, and costs for trade publications.

6. Cash budget. This budget shows the source and amount of income and expenditures on a monthly basis. Its purpose is to determine how much cash will be needed and when, so that excess cash can be invested. It is not uncommon for large corporations to earn $3 million or more a month by investing their unused capital for periods of 30 to 90 days. Conversely, if there is a shortage of funds, then the firm must liquidate assets or secure short-term loans to pay its current bills. Since many of these obligations are to members of the supply chain, the terms and conditions developed by purchasing can have a significant effect on the firm's cash flow statement.

PRINCIPLES OF BUDGETING

Decentralized Budgeting

The manager charged with responsibility for controlling department expenses should have the authority to propose the budget for that department. Many large organizations do not adhere to this principle but centralize all budgeting, having the expense budget prepared by someone other than the responsible manager. Some organizations have a budget section in their accounting department which prepares budgets for all departments. Other companies have the budgets developed by upper middle management and top management. Four primary reasons support the use of decentralized budgeting.

1. Removing the manager's authority to prepare his or her own budget violates the principle of equality of authority and responsibility. If top management tells a manager, "You are responsible for your area, but we will tell you how much money you can spend, what you can spend it on, and when you can spend it," then his or her authority is undermined. Allocation of funds determines in large part how a department will be run; to remove budgeting responsibility from a manager severely weakens his or her ability to perform effectively.

Managers do not, however, need to have absolute authority over their budgets. They should be given the responsibility for initiating the budget request. Nevertheless, the budget, like most other matters, should be coordinated through the chain of command for approval at each level. In other words, managers should be able to propose budgets—not dictate them.

2. The second reason for decentralized budgeting is that it is motivational to the manager involved. Real participation in management decision making is one of the best ways to achieve true motivation. The decentralized budget is an excellent tool for getting managers at all levels involved in decision making. Decentralized budgeting is motivational in another way. When managers are responsible for preparing their budgets and also for enacting them, they have a vested interest in seeing their plans succeed. They have put their reputations on the line by submitting their budgets. In effect, they have said, "If you approve this budget, I will meet it for you."

3. The third advantage of decentralized budgeting is that it encourages managers to plan for the coming year. The preparation of annual expense budgets forces managers to analyze what they are going to do, expense item by expense item, for an entire year.

4. Finally, decentralized budgeting makes it necessary for the responsible managers to understand what items make up the department expenses. This knowledge is helpful in controlling costs.

Budgeting Accountability

Managers should be held accountable only for costs over which they have control and not for costs over which they have no control. Many of the normal costs of doing business are controllable by someone. This is not to say that every cost is controllable in every department. For example, purchasing managers cannot control the price of steel or whether an individual production department needs steel. In the same sense, store managers can hardly be held accountable for a general increase in wages of 50 cents per hour that is negotiated by the corporate labor relations staff. However, they should be accountable for letting stock clerks take four hours to complete a two-hour task—and thereby increasing total labor cost. According to this principle, each cost should be identified with a manager who can control that cost. In a few cases, it may not be possible to identify a cost with just one manager. In these situations, one must use common sense to work out the best compromise possible under the circumstances. For example, overhead charged to each department as a percentage of monthly sales is not a controllable item for a production department.

To know how much it costs to make a certain product or service or to run a purchasing department, it is necessary to allocate overhead costs. Traditionally, overhead costs have been allocated based on direct labor hours or dollars. The problem with such an approach is that the proportion of direct labor as a percentage of total costs has declined dramatically in recent years. Unfortunately, such conventional allocation methods often lead to poor decisions based on misrepresenting the actual costs incurred. To help alleviate these problems, activity-based costing techniques have been developed to reflect more accurately the actual proportions of overhead used in particular activities. In this context, cost drivers, or causal factors, are identified and are used as the basis for allocating overhead.

Activity-based costing uses a two-stage allocation process and starts by assigning overhead costs to cost activity pools. These pools represent ongoing activities such as obtaining or comparing quotations. In the second stage, costs are assigned from these pools to activities based on the number or amount of pool-related activities required to complete them. In this context, transactions become extremely important, and low volume (or value) activities that necessitate proportionately higher transactions (or purchasing involvement) necessarily warrant higher costs. The message for purchasing professionals is clear: avoid those activities that consume proportionately higher time or involve inordinately high numbers of transactions unless they can be costed at a higher rate. Becoming familiar with these techniques also enables purchasing to use a transaction focus for enhanced traceability of overhead costs and ultimately more accurate unit cost data for purchases.[1]

Reporting Budget Results

Managers should be required to make regular reports to their superiors that explain the reasons for significant variations between actual and budget expenses. There are two reasons for this principle. The first is to keep superiors informed about deviations from plans and the reasons for them. Suppose, for example, that a manager has a $4,000 surplus in the staffing account for the month of May. Why is there a surplus, and what are its implications? Perhaps the surplus exists because the addition of a new buyer scheduled for the month of May had to be postponed. Thus, the salary saving is affected by specific work not done, and the information provided by the budget report can be critical to others in the organization.

The second reason for requiring responsible managers to make regular budget reports is to provide them with an incentive. Preparing such reports requires managers to investigate and analyze the reasons for variations in the spending plan. Knowledge of departmental costs and what affects them leads to better control of those costs.

METHODS OF BUDGETING

Zero-based Budgets

Zero-based budgeting is a process that *does not* use past experience to determine future needs. All budget items are fully justified in detail and are viewed as *new* requests as opposed to continuations of current programs. The assumption is that the budget is prepared from scratch (zero based). The zero-based budget is helpful in questioning traditional practices

---

[1]For additional information on activity-based costing, please see Chase, Richard B. and Nicholas J. Aquilano, *Production & Operations Management: A Life Cycle Approach,* Sixth Edition, Homewood, IL: Irwin, 1992; Johnson, Thomas and Robert Kaplan, *Relevance Lost: The Rise and Fall of Management Accounting,* Boston: Harvard Business School Press, 1987, and Garrison, Ray, *Managerial Accounting,* Homewood, IL: Irwin, 1991.

because all programs, including those that have been in effect for years, are justified, prioritized, and subjected to scrutiny and approval.

In reality, few organizations use a "pure" zero-based budget. Those that do generally employ the basic premise for selected operations and use the traditional "historical/extension" concept for developing the budget for the rest of the operation.

Line Item

A line item budget (which is the most commonly used) is formatted to show individual expenses during the budgetary period, without tying those expenses into broad programs or goals. A typical line item budget is divided into such categories as salaries, office supplies, travel, equipment, telephone expense, and postage (Table 5.1). Each of these categories contains further detail of what these expenses are. As an example, the travel section shows exactly what travel is to be taken and how much each trip is estimated to cost. Line item budgets are generally incremental, meaning they are to a large extent based on the previous budget period.

Program

Program budgets often use Program Planning Budgeting Systems (PPBS). This type of budget is often used by nonprofit and governmental entities. Program budgets link the organization's goals and objectives with the programs or sections responsible for meeting those objectives. To further this relationship between goals and expended funds, this type of budget normally uses productivity measurements and cost benefit analysis. Program budgets have the advantage of offering management the opportunity to evaluate and make decisions on the need for various programs.

Flexible

Flexible budgets change depending on changing conditions, such as an increase or decrease in output. One type of flexible budget is known as the variable budget, which is a set of budgets that vary to account for different conditions. Often, flexible budgets use a formula to determine the requisite budget amounts based on the output. The obvious advantage of flexible budgets is that they respond quickly to change.

TABLE 5.2
Budgeting Steps

Budgeting Steps
Review goals & objectives
Define needed resources
Estimate the dollar value of needed resources
Present the budget/obtain the appropriation
Control expenditures

THE BUDGETING PROCESS

Table 5.2 lists the steps necessary in budgeting.

Review Goals and Objectives

The first step in the budget process is to review budget unit goals and objectives to ensure congruence with organizational goals and objectives. This is important so that the budget demonstrates in financial terms how these goals and objectives are to be met.

Define Needed Resources

The next step is defining the needed resources in the budget process. It is necessary to begin with general forecasts in terms of economic trends, purchase prices, sales, and profit. This step requires realistic figures for revenues and expenditures. While some or all forecasts may be provided by top management, finance, or marketing, it is generally agreed that the actual budget requests are best developed at the level where implementation takes place (usually at the department level or lower). This approach tends to work best because those responsible for implementation can best identify their own needs, and they will be better motivated if they have input into the decision-making process.

Estimate the Dollar Value of Needed Resources

Clearly, the dollar value of needed resources must be estimated accurately for budgets to be valid. The best starting point for this step is to estimate values by closely analyzing the previous year's actual expenditures. This information can then be used to extrapolate resource figures for the new budget year.

Present the Budget/Obtain the Appropriation

This step is handled differently depending on the organization. It is quite common to establish a committee to review all budgets, but there are many other approaches used. After the budget has been presented and any changes have been made, appropriations are adopted to cover the approved expenses during the budgetary period.

Control Expenditures

The final step in the budget process is the control of expenditures during the budgetary year. The budget is the most widely used tool in organizations to provide financial control. This control occurs through the matching of appropriations and expenditures and also by tracking expenditure trends against budget estimates. It is essential to document the lessons learned during these comparisons so that good practices can be repeated and less satisfactory techniques can be avoided.

COST CONCEPTS

A budget is a formal written statement that expresses planned future operations in financial or numerical terms. To be effective, budgets must contain some means by which management can determine whether planned operations are being accomplished. As such, there are two elements common to all budgets. *First, there is a set of specific goals that relate to future operations*. The establishment of goals is equivalent to defining the standards by which the firm is to measure its performance. *Second, provision is made for a periodic comparison of actual results and established goals.* This represents the control feature of budgeting activity and is usually accomplished through the development and use of budget performance reports.

With few exceptions, most budgets are developed in terms of *cost.* In fact, one of the major duties of the purchasing manager is to review and evaluate a potential supplier's actual or anticipated costs. Evaluation itself involves the judicious application of experience, knowledge, and judgment to the seller's cost data. The purpose of this evaluation is to project reasonable estimates of contract costs. These estimates then become the basis for negotiations between buyer and seller, to be used for arriving at contract prices mutually satisfactory to both parties.

Standard Costs

Standard costs can be defined in terms of either manufactured items or units of service:

1. ***Manufacturing***—Standard costs are the *predetermined* costs of manufacturing a single unit or a number of product units during a specific period in the immediate future. These costs are the planned costs of a product under current and/or future expected operating conditions.
2. ***Service***—Standard costs are the *predetermined costs* of providing a single unit of service or a number of service units during a specific period in the immediate future. These costs are the planned costs of providing a service under current and/or future expected operating conditions.

It should be noted that, regardless of the area, standard costs are carefully predetermined. Accounting departments frequently ask purchasing to participate in estimating the standard costs for materials to be used during the coming budget period. These are costs that should be attained under efficient operations. As such they are the costs used as a norm for the product or service and become the goal or benchmark against which actual costs will be compared in developing cost variance reports.

Proper use of standard costs can be of great benefit to management. For the purchasing manager, they provide a basis for developing budgets, controlling costs, and measuring efficiencies. Standard costs are also used in the reduction of costs, and in establishing bids and contracts.

Direct Costs

Direct costs are those that can be identified with specific products or services. These costs accrue from the unit being produced and, as a rule, are classified as being either direct *labor* cost, direct *materials* cost, or purchase costs. *Direct costs are always treated as variable costs.* (This means that direct costs are 100 percent variable.) With the exception of those industries in which there are large investments in fixed capital, direct costs generally represent the greatest percentage of total costs. This fact is of special importance for the following reasons:

1. Direct costs generally serve as the basis from which sellers make their allocation of overhead costs.
2. A reduction in the direct costs of a seller is generally worth more to the buyer than a major reduction in the seller's profit percentage. This is an important factor to be utilized in the negotiation process.

Indirect Costs

Indirect costs are those that can be identified with the operations of the organization (e.g., production process) but not with specific products or services. As a general rule, indirect costs are all those that are not classified as direct costs. There are three basic categories of indirect costs: *fixed*, *variable*, and *semivariable*.

1. **Fixed costs** represent costs that tend to remain constant, regardless of the volume of operating activity. In particular, a fixed cost meets the following criteria:
   a. The cost remains constant within a defined range of operational activity.
   b. The cost decreases as a cost per unit (i.e., average cost decreases) when output levels are increased.
   c. The cost is assigned to departments by either managerial decision or an appropriate cost allocation method.
   d. Incurring the cost is a function of top-level management, not lower-level supervisors.
2. **Variable costs** are expected to fluctuate in direct proportion to changes in the level of operational activity (e.g., sales, production levels, or some other measure of activity). Variable costs are

constant on a unit basis, but tend to vary in direct proportion to changes in volume when measured over a specified period of time. In particular, a variable cost meets the following criteria:

a. The cost exhibits a variability of total amount that is in direct proportion to changes in the levels of operational activity.
b. The cost per unit is relatively constant even with changes in the levels of operational activity.
c. The cost can be assigned to operating departments in an easy and reasonably accurate manner.
d. Consuming and incurring costs are controlled by the department head or manager who is responsible for them.

3. **Semivariable costs** display both fixed and variable characteristics. Examples of semivariable costs include salaries of supervisors and buyers, pension plans, utilities, and fuel. Such costs must be analyzed to determine both the fixed portion and the variable portion. This is generally accomplished by direct application of one of three historical approaches (the method of high and low points, the method of least squares, and the method of visual analysis of a scattergram) or by use of an appropriate analytical approach. As a general rule, a semi-variable cost is one that:

   a. Tends to change in proportion to changes in the level of operational activity, but not in direct proportion
   b. Can be separated into a fixed and a variable element (e.g., a buyer's salary that consists of a fixed monthly rate plus a bonus, which is the variable portion)

COST SAVINGS

Cost savings are multifaceted and can be attempted by individuals or by areas, departments, divisions or corporate-wide programs. Because of this flexibility, a purchasing manager does not need to wait for direction to initiate a cost-savings program. To cut costs: (1) adopt an open-minded attitude; (2) choose a cost to be reduced; (3) identify target areas; (4) motivate your staff and get their ideas; and (5) question every element of the costs to be reduced.

To start a cost-savings program it is necessary to cultivate the proper mental attitude. In today's downsizing environment, cost reduction equals job loss in many people's minds. The ideal situation is to get agreement that people will not be eliminated when these programs result in position changes or reductions. Instead they should be trained for jobs in other parts of the organization. Even the discussion of a cost reduction drive predisposes the wrong mental attitude. It may be necessary to substitute a title such as "cost-savings program," or internal maximization of profit (IMP) or some other less distasteful phrase. The long-term goal should be a never-ending attention to cost savings instead of the traditional cost reduction drive that top management adopts during a profit squeeze and forgets about when profits improve. An effective cost-savings program should be different and should maintain a constant, realistic, continuing pressure to save money regardless of economic cycles.

The second step is to focus on specific costs to be reduced. The budget provides a good starting point and describes where most of the money is being spent. Often, it is wise to target the large ticket items first. For example, if you spend $30,000 on one material and $1,000 on supplies each month, you have a much better chance to make a large savings in the one material account than in the supplies account.

As the third step, identify two or three specific cost items as targets. During a period of several months, it is wise to take these items apart and scrutinize them closely. The idea is to zero in on these items—to know them, think about them, and analyze them continuously. Since much of a manager's job is getting work done through people, it is necessary to develop and cultivate a strong desire in subordinates to support the cost-savings efforts. This is the fourth step in the cost-savings process. One of the best approaches in motivating employees to help reduce costs is to solicit employee participation. It is useful to convene a departmental meeting to explain the importance of cost savings to them and to the firm. The proposed effort should not be a crash program but rather a reasonable and long-term effort. The help and suggestions of employees are expected and are critical to the success of any cost-savings initiatives that might be implemented. It is necessary to specifically ask for their ideas and comments and to contact them individually during the days following the departmental meeting to get their suggestions. Above all, they have to truly believe that the effort is important to the organization and to the purchasing leader, so it has to be continually reinforced over a period of time. The purpose of visiting with each person in the department is twofold: (1) to make a personal appeal for their suggestions; and (2) to get them to become cost conscious about their particular jobs.

It is important to remember that each person usually spends about 40 hours a week doing a job and thinking about it. For this reason employees are likely to know more about their jobs than anyone else. Many times they have some good ideas about how improvements can be made but are never asked for suggestions. Or even worse, when they make suggestions that are delayed or never implemented, employees often start seeking their own agendas.

Similarly, it is also worthwhile to involve other members of the supply team, especially suppliers, internal customers, and even end users in cost reduction efforts. Many organizations have achieved significant cost savings by capitalizing on these types of relationships.

As the fifth step, every element of each job, material, or machine involved in the accounts that have been chosen for cost reduction should be scrutinized. It may be helpful to ask the following questions:

1. Why is it necessary? Could it be eliminated?
2. What is its function? Can it be accomplished in a better way?
3. Who should do it? Is the right person doing the task now and is that person properly trained and motivated?
4. When should it be done? Can cost be reduced or efficiency increased by changing the time when it is done?
5. Where should it be done? Does location affect cost or efficiency in any way?
6. How should it be done? What method, materials, or machine could reduce costs?

The objective of this questioning should be to break each item down to the smallest unit possible. Cost reduction ideas usually occur based on a critical understanding of the individual parts that make up the whole. It is important, however, to ensure that cost reductions for individual facets of a task or function do actually contribute to overall cost reductions. For example, a cost savings attributed to the purchase of lower quality parts for a subassembly that subsequently results in higher rework costs cannot truly be considered a savings.

Value Analysis

Value analysis (VA) is a systematic approach to finding unnecessary costs in a product or service and eliminating them. The goal is to increase profit by analyzing and improving the functions of goods and services for the lowest life-cycle cost. VA is a study of the function of the item being used. Automobile manufacturers found, for example, that they could save millions of dollars by changing the overflow pipe on automobile radiators from copper to rubber. They simply asked, "What is the cheapest way to remove overflow of radiator fluid?" Answer—rubber tubing. "Then why are we using copper—one of the most expensive metals?" It was a good question, and it takes many good questions about things occurring throughout the supply chain to bring about cost savings.

The principles of VA can be useful to anyone involved in the operations of a company, but they are particularly relevant to purchasing as the managers of external manufacturing. VA is based on two premises: (1) all products and services have necessary and unnecessary costs; and (2) unnecessary costs can be eliminated by applying the methods of VA, which coordinates all functions within the organization in an effort to reduce the cost of producing and selling the product or service. A simplified VA procedural checklist is provided in Table 5.3.[2]

In the context of VA, it is important to remember that value can be interpreted differently depending on an individual's perspective. The definitions of value often considered are as follows:

1. *Use value* includes the properties and qualities that allow the product to accomplish its intended function.
2. *Esteem value* refers to pride of ownership and relates to the "sell" function of a product.
3. *Cost value* is the sum of all labor, materials and overhead expenses required to produce the product.
4. *Exchange value* represents the properties or qualities of an item that enable it to be traded for something else.

In addition, the mathematical definition of value is:

worth/cost

In other words: if the value is > 1, it is good

= 1, it is average value

< 1, it is poor

[2]For additional information on Value Analysis as it applies to purchasing, please view the NAPM tape, "Principles of Value Analysis" (Program Aids Library #20), produced in 1989.

TABLE 5.3
VA Methodology

VA Methodology
Consider the contribution of design, materials, equipment, & personnel to product value
Determine whether the cost of any area is proportionate to its usefulness
Evaluate the need for each feature in a product, component, or equipment against the user's expectations
Search for better designs, materials, tooling, processes & procedures
Promote standardization among the organization's products & processes
Evaluate whether the cost elements accurately reflect the true cost of the product
Explore whether a product or service can be purchased at lower cost
Investigate whether other organizations are buying the same item for less

Obstacles to VA and basic reasons why a product may contain unnecessary costs are shown in Table 5.4. Focusing on a product's function rather than its design sometimes minimizes these obstacles and the friction between departments within an organization. VA can be done on a formal or informal basis, and it usually consists of six phases (shown in Table 5.5). The initial step before beginning a VA Job Plan is to identify the function a product performs. A verb-noun definition is used (e.g., the picture tube displays images). Function is also evaluated by comparing the total cost of an item to the worth of the function provided. A dollar figure must be assigned to establish the worth.

Applying the VA Job Plan requires objectivity, selectivity, simplification, and allocation. Teamwork is essential and often includes members of the supply chain. Information is shared across departmental and frequently, organizational boundaries. VA improves reliability, since all aspects of the

TABLE 5.4
Obstacles to VA

Obstacles to VA
Changing market conditions
Lack of information
False information
Lack of time
Calculation shortfalls
Idea inhibition
Resistance to change
Traditional habits and attitudes
Trade-offs required to meet schedules

product are examined and allocating resources effectively (often using a decision matrix) brings reality to VA.

Often, through value analysis, cost reductions of 15 to 25 percent can be achieved without reducing the value of the product or service. A survey by *Purchasing* magazine found that the average payback on value analysis programs was $26 for every $1 spent to reduce cost. Black and Decker Manufacturing Company reported a ratio of 100:1 savings to cost or $1.2 million savings on a program that cost $12,000. Potential VA candidates are shown in Table 5.6. In the final analysis, it is advisable to apply a VA checklist and to adopt functional cost techniques. Brainstorming among members of the supply chain, and the use of supplier seminars and demonstrations for detailed cost analyses, are also useful. Again, it is important

TABLE 5.5
Value Analysis Job Plan

Value Analysis Job Plan
Information phase
Speculation phase
Analysis phase
Development phase
Execution phase
Evaluation phase

TABLE 5.6
Potential Value Analysis Candidates

Potential Value Analysis Candidates
Raw materials
Component parts
Production tools
Electrical equipment/supplies
Materials handling equipment
Packaging & shipping
Office equipment & supplies
MRO & safety supplies

that the purchasing professional views his or her role as one of managing external manufacturing rather than buying purchased items. With this perspective in mind, it is possible to achieve significant breakthroughs in cost reductions and, more important, value enhancements. While nearly everyone believes that value analysis is a good way to reduce cost, very few organizations actually make a serious effort to teach employees the techniques of value analysis. By establishing a good value analysis program, purchasing managers can play an important role in their organization's overall cost reduction efforts.

INVENTORY CARRYING COST

Another fertile area for cost reduction is inventory carrying costs, which are made up of the following:

1. *Insurance Costs*—Insurance cost to insure the inventory against losses from fire, theft, or water damage runs from 1 to 4 percent of the inventory's value.
2. *Management Cost*—The cost of keeping records and controlling the inventory is about 1 percent of the value.
3. *Obsolescence, etc.*—Obsolescence, damage, and theft can run from 1 percent up to 15 or even 20 percent in some types of businesses.
4. *Property Taxes*—In many states, taxes (often as much as 4 percent) are levied against the value of business assets. For a manufacturer, inventory is frequently the largest asset they own.

A purchasing manager should establish a sound cost reduction program and do the following:

1. Provide training for those involved
2. Help set individual and departmental objectives for cost reduction
3. Make resources available as needed
4. Provide leadership and help with cutting red tape, selling change, etc.
5. Have everyone report monthly on this cost savings
6. Produce and circulate an annual organizational report on cost savings
7. Give recognition and rewards

In top purchasing organizations, members of the supply team work jointly to use standardized products, sub-assemblies, and parts to minimize costly customized design as well as inventory investment. Additionally, inventory investment is optimized throughout the supply chain. The use of an inventory and purchasing budget allows inventory monitoring and timely, closely controlled releases of PO requirements.

KEY POINTS

1. Effective budgeting and cost reduction are critical to the success of any purchasing organization. Financial plans are used to project income, costs, and profits in the near future. Budgets showing sources and allocations of funds usually accompany these plans.
2. A budget is used to allocate resources, control expenses, and measure performance. To control and reduce costs, it is necessary to first know what items constitute total costs, how much should be spent, and how much is being spent. In addition, the process of budgeting and expense comparison is a cost-savings, technique in itself, because paying attention to costs inevitably causes them to go down.
3. Revenue budgets forecast income for the coming year. A forecast's requirements provide the basis for determining materials, expense, capital, and cash budgets.
4. Materials budgets project the amount and costs of supplies (raw materials or goods) that need to be purchased in the next budget period. It helps the purchasing manager establish a purchasing schedule and estimate purchasing's financial requirements.
5. The capital budget specifies the amount of money to be spent for plant and equipment and is important because of the magnitude of funds involved and the longer budget time frames.
6. The expense budget specifies the funds required for operational and administrative tasks, including salaries and overhead.
7. The cash budget shows the balance between cash inflows and expenditures, and determines the cash/financing needs of the organization on a month-by-month, or week-by-week basis.
8. Decentralized budgeting ensures that a manager who is responsible for controlling department expenses has the authority to propose the department's budget.

9. Managers should be responsible for budgetary items for which they have control. Activity-based costing ensures that overhead costs are allocated fairly and accurately.
10. Managers should write regular budget status reports to supervisors to provide incentive for the managers and to ensure the maximum benefit gained from available resources
11. Zero-based budgets are not based on historical data and are useful in justifying the need for current activities.
12. A line item budget shows individual expenses during the period and is generally based on previous budget periods.
13. In preparing a budget, a purchasing manager should review the organization's goals and strategies, define needed resources, estimate their value, obtain the necessary funds, and control or monitor the related expenses during the budget's period.
14. Budgets are developed in terms of costs, and purchasing managers are required to determine reasonable estimates of these costs.
15. Standard costs are carefully predetermined manufacturing or service costs. Direct costs are costs that can be linked to a specific product or service. Indirect costs are those identified with the operations of the organization.
16. Value analysis (VA) is a systematic way of finding unnecessary costs in a product or service and eliminating them. The goal is to increase profits by analyzing and improving the functions of goods and services for the lowest life-cycle cost.

REFERENCES AND RECOMMENDED READINGS

Leenders, Michiel R., Harold E. Fearon, and Wilbur B. England, *Purchasing and Materials Management,* Tenth Edition, Homewood, IL: Irwin, 1993.

CHAPTER 6

ORGANIZING AND ORGANIZATION

FUNDAMENTAL IDEAS ABOUT ORGANIZING

During the presidential campaign of 1840, Abraham Lincoln wrote a circular designed to help organize the Whig Party in Illinois to get out the vote for Harrison. He wrote, in part:

> the duties required of each county committee: To divide their county into small districts and to appoint in each a subcommittee, whose duty it shall be to make a perfect list of all the voters in their respective districts, and to ascertain with certainty for whom they will vote...on election days see that every Whig is brought to the polls.[1]

Lincoln knew what the Greek and Roman generals had learned 2,000 years before; namely, that organizing and organizational structure are basic to success.

As Fayol stated, "To organize a business is to provide it with everything useful to its functioning: raw materials, tools, capital, personnel. All this may be divided into two main sections, the material organization and the human organization."[2] Every manager needs to know how his or her department fits into the overall organizational scheme. It is also helpful to know what staff and service personnel are available, as well as what the requirements and limitations are. The company organization chart is the key to understanding these relationships.

---

[1]Lopsley, Arthur Brock, ed., *The Writings of Abraham Lincoln,* Vol. 1, New York: G.P. Putnam's Sons, 1923, p. 230.

[2]Fayol, Henri, *General and Industrial Management,* Sir Isaac Pitman and Sons Ltd., London, 1969, p. 53.

Basic Concepts of Organizing

Before exploring the various aspects of organizing, some fundamental concepts should be defined. Chapter 4 illustrates that organizational goals and plans lead to organizing and the organizational structure. The basic concepts of organizing include the following:

1. Organizing is the work of providing, in advance, those things needed to carry out a plan. It is the process of assessing the requisite organizational activities, designing tasks, grouping tasks in manageable units and establishing the relationships between tasks and groups of tasks.
2. An organization is a group of people working together, under the direction of a leader, to enact certain plans.
3. Organizational structure is the framework of relationships between the parts of the organization.
4. Organization principles are guidelines that make organizational structure workable.

This chapter discusses the function of organizing in general, as well as several aspects of organizational structure and the principles of organizational behavior.

Purchasing managers are not necessarily responsible for the overall organizational structure, but they must often organize a department to carry out specific plans or programs. As shown in Figure 6-1, organizing involves four basic steps. Each manager must first translate broad corporate goals into more specific unit objectives. Subsequently, he or she must determine the activities necessary to achieve organizational aims, which often involves determining the sequence and timing of necessary activities and tasks. Next, the manager must arrange the activities into logical groupings or work units that can be performed by individuals. Finally, he or she must assign an individual the authority and responsibility for each array of tasks to ensure their successful completion.[3]

Managers must reorganize whenever a change takes place in a business's objectives or policies, as organization is the first step in putting any plan into action. Time and money are wasted because managers fail to

---

[3]Adapted from Badawy, M.K., *Developing Managerial Skills in Engineers and Scientists: Succeeding as a Technical Manager,* New York: Van Nostrand, 1982, p. 154.

FIGURE 6-1
Organizing

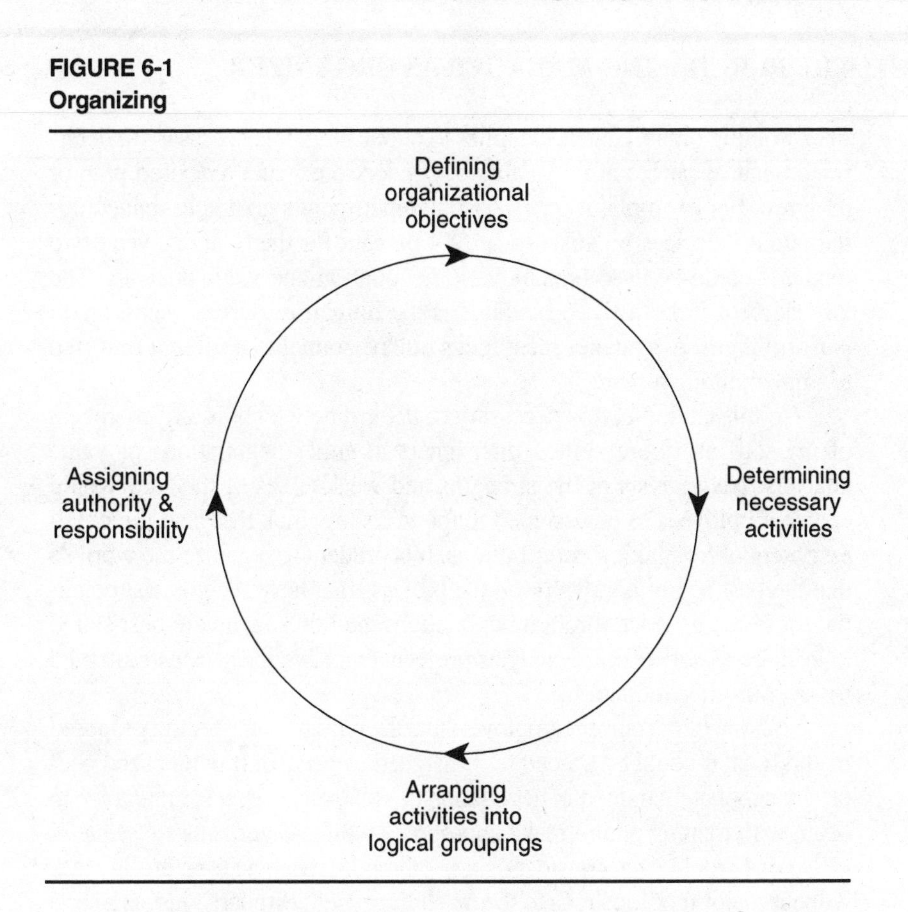

organize properly. To prevent this type of waste, the following actions, although not all-inclusive, are necessary:

1. Make the right people available for the project.
2. Ensure that these people know their roles in the organization and how their jobs relate to others.
3. Confirm that these people know specifically for what part of the plan they are responsible.
4. Assure that they are properly trained to carry out their part of the plan.
5. Guarantee that they have the resources they need (equipment, building, and materials) at the right time and in the right place to carry out the plan.

THE PURCHASING MANAGER AS ORGANIZER

In examining organizing as it applies to purchasing, it is necessary to determine what resources are available or needed to execute a desired plan or program. For example, if a purchasing department's goal is to reduce initial purchase prices by an average of 4 percent for the financial year, two methods could be used to achieve it, negotiation and value analysis. The key element in both these methods is the human resources necessary to perform them. A manager must focus on "personnel issues" as a first step to implementing a plan.

In this example, it is necessary to determine which, if any, members of the staff are more skilled than others at either negotiations or value analysis. An analysis of the strengths and weaknesses of the staff would enable employees to be assigned to the tasks in which they had the greatest potential for making contributions. It is prudent to organize the work so that the best negotiators work on the projects that have the greatest potential for reducing price through negotiations, and those who are best suited to lead are assigned value analysis projects that have the greatest potential for overall price reductions.

If none of the current employees fits the requirements for the proposed methods, staff could be trained or reassigned as needed. If none of the people are adequately trained in negotiation or value analysis, it is imperative to begin with training and to make appropriate work assignments subsequently. No one would even consider using a stamping machine for a drilling task without major modifications to the machinery itself. Similarly, just as a shop supervisor should not ask a pipe fitter to do a welder's job without retraining, a purchasing manager should not ask employees to undertake new tasks without investing in the training and development they need. An investment in developing purchasing professionals is just as important before asking them to undertake new tasks.

To complete organizing the department, it is necessary to analyze each facet of the overall plan from the perspective of what is required to make each plan work. Does the organization need more people? Can it achieve the same results with better training and development of existing people or does it require more motivated people? Can the organization achieve its goals with enhanced computer, facsimile, or EDI capabilities? Does the organization require more (or sometimes less) support from top management? Or can the organization leverage its supply chain by garnering better support from the other departments? Depending upon the answers to these questions, it is the purchasing manager's job to obtain the resources required to enact the plan.

In addition to the organizing tasks discussed above, a purchasing manager must also be concerned about the organizational structure and its operations and staffing. The remainder of this chapter addresses the question of structure, while staffing is discussed in the next chapter.

COMPONENTS OF ORGANIZATIONAL STRUCTURE

In the company as a whole, organizational designs are often quite varied. Figure 6-2 shows the organizational structure of a manufacturing company. However, no matter how complex an organization may appear, it usually contains only three organizational components or building blocks. They are (1) line organization, (2) staff organization, and (3) service organization (Figure 6-3).

Line Organization

The line exercises the organization's primary functions. For example, in a company that manufactures lawn sprinklers, those departments that produce the sprinklers are line departments. In a wholesale meat house where the primary function is the sale of meat, the sales force is the line department. The primary function of an army is to wage war, so the combat units are considered line organizations. As organizations grow larger and more complex, the line departments usually need assistance. They obtain it from staff and service departments, which take over some of the duties previously performed by the line organization.

The Staff Organization

The term *staff,* in the past, meant a stick held in the hand for support. In an organization, *the staff provides support by giving advice to members of line, service, and other departments.* The staff may also have functional authority and make functional decisions. In manufacturing and retailing, for example, the legal department and the labor relations department are staff functions since their primary duties are advisory. Obviously, the number of advisers that a company can afford to have is relatively small. This fact accounts for the scarcity of staff managers, particularly at the lower levels of the organization.

FIGURE 6-2
Manufacturing Organization Chart

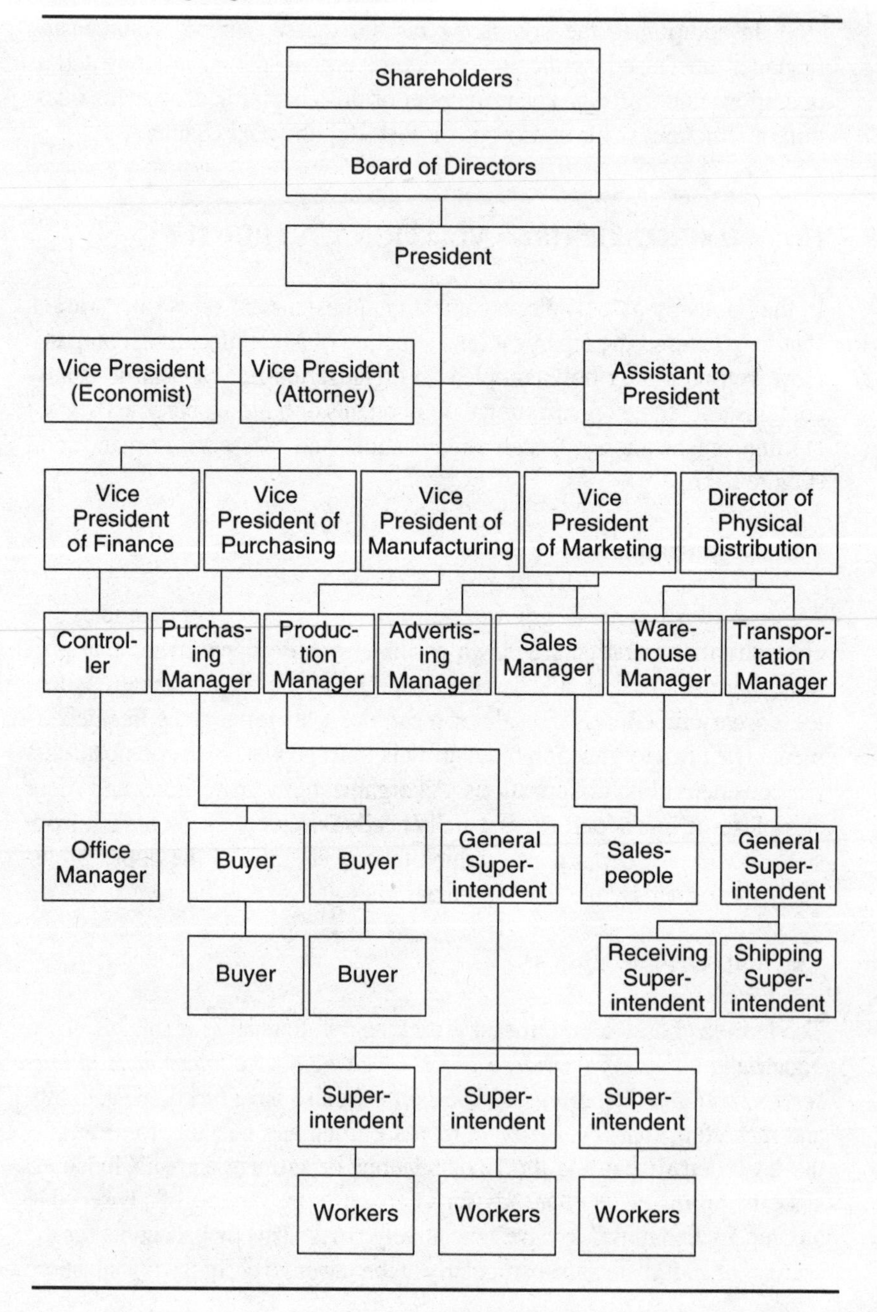

FIGURE 6-3
Organizational Structure Components

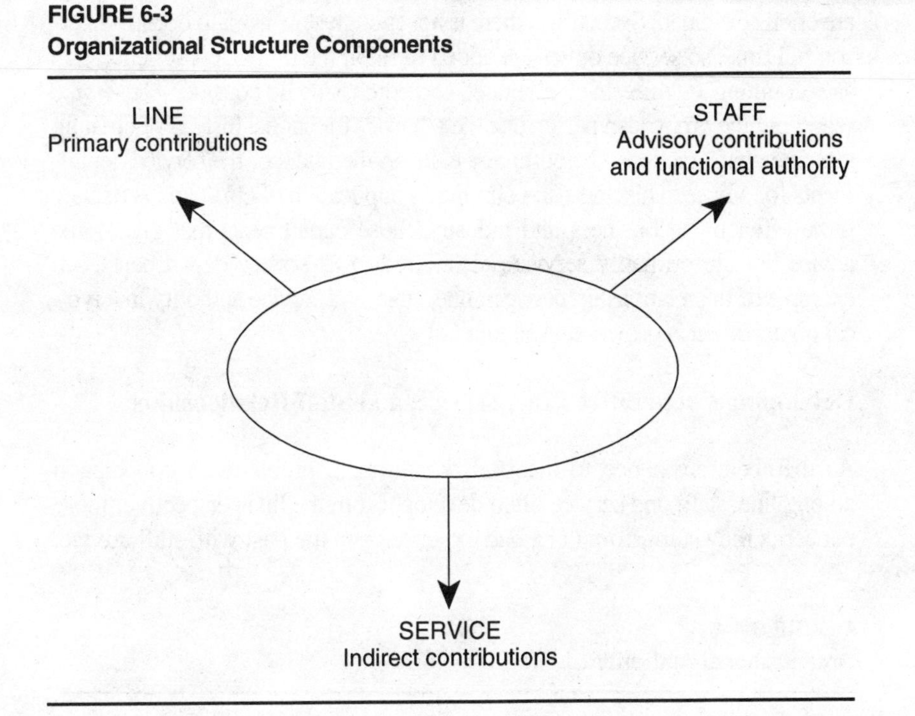

The Service Organization

The third part of an organization is service. Many texts have underemphasized service by combining the functions of staff and service and calling them both staff. *A service department is one that makes an indirect contribution to the achievement of the organization's primary function.* To illustrate, in manufacturing and retailing, the accounting department employees do not actually produce a product, but they aid the line operations. If there were no accounting department, line managers would have to prepare the payroll for their departments. The supervisors would have to make the appropriate deductions and then write each employee's paycheck. In addition, they would need to keep records of how much was paid out to various suppliers and why. They would also have to file tax reports with the government. The accounting department relieves the line managers of these and other responsibilities.

In practice, the demarcation between staff and service is seldom distinct, especially at the lower levels of an organization where these functions are often combined. Sometimes there is not enough staff work to occupy a person full time, so service duties are added to fill in the time. Departments such as accounting, engineering, personnel, and purchasing do considerable service work, but they may also perform some advisory functions. Instead of creating separate staff functions, the company calls on members of the service departments for advice. This approach eliminates duplication of effort and is usually more efficient. In business and industry, those departments that give some advice but are primarily service are referred to as service departments. An example of the relationship between line, staff, and service authority in a typical organization is provided in Figure 6-4.

Developing Cooperative Line, Service, and Staff Relationships

As members are added to the staff organization, problems of cooperation among line, staff, and service often develop. From the line perspective, these concerns may stem from (1) a bad experience in the past with staff advice,

FIGURE 6-4
Organizational Authority

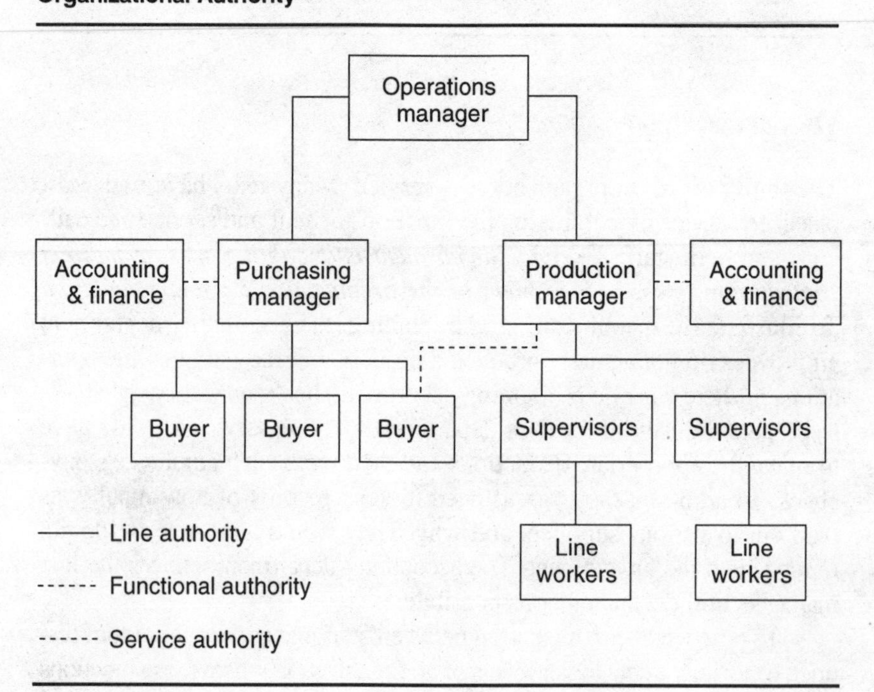

(2) a dislike for leaving decision making to staff members who will not have to assume full responsibility for those decisions or (3) the belief that asking for advice is a sign of weakness.

Staff, on the other hand, may fail to cooperate because of (1) an inability to sell its ideas, (2) a preference for dictating instead of advising, or (3) failure to fully understand line and service positions. While these problems can never be completely overcome, they can be minimized if staff members follow these suggestions:

1. Work with and through the managers they are trying to help, instead of trying to go around them.
2. Actively seek the line or service manager's ideas and cooperation, while showing respect for his or her knowledge and experience.
3. Give advice in the form of suggestions rather than orders.
4. Share credit for advice that turns out to be good with those who helped to put it into practice, instead of taking full credit themselves.
5. Accept responsibility for giving bad advice rather than blaming others for poor execution.

Line and service managers who need staff assistance should remember to:

1. Ask for help before a problem becomes critical.
2. Leverage, the experience gained by the experts instead of wasting time "reinventing the wheel."
3. Provide feedback to staff on the results their advice.
4. Share recognition with staff for advice that resulted in success.
5. Accept responsibility for poor execution of good advice instead of laying the blame on staff.
6. Realize the interdependence between staff and line or service.

DISTINCTIONS BETWEEN LINE AUTHORITY AND FUNCTIONAL AUTHORITY

In the study of organization, people are often confused by the dual usage of the word *line*. *Line organization* refers to the departments that carry out the organization's primary function. But when the word *line* is used in connection with *authority,* it means something quite different. ***Line authority*** *is the right and the responsibility to command others.* All departments, whether they are line, staff, or service, have managers with line authority. Workers usually think of these managers as supervisors.

The purchasing department of a manufacturing firm provides a good example of line authority in a service department. The purchasing manager may have five buyers and ten junior buyers responsible to him or her. In this instance, the purchasing manager has line authority over them and is in fact their supervisor.

Functional authority is the authority for a line, staff, or service manager to make decisions affecting employees and/or operations that are under the direct control of another manager. For instance, quality control managers may be authorized to shut down a production line making transistors. Their authority to make shutdowns may be stated as follows: "Quality control managers are authorized to stop production on any line where more than 0.6 percent of the transistors being produced are defective." The statement of authority adds that neither department foremen, general foremen, nor even production managers can override the decision of the quality control manager.

Whenever functional authority is used, it violates a very sound principle of management (the principle of accountability to a single superior). Because it breaches a principle of management, functional authority must be used judiciously. To use it well, everyone involved should understand (1) who has functional authority over what and/or whom, (2) why this functional authority is necessary, and (3) what the boundaries of authority are, as in the example of the 0.6 percent defective transistors mentioned above. Recently, many organizations have replaced traditional functional authority with cross-departmental work teams to enhance interdepartmental cooperation and communication. The role of these teams is discussed later in this chapter.

CENTRALIZED VERSUS DECENTRALIZED CORPORATE STRUCTURE

An important topic in the study of organizational structure is the difference between *centralized* and *decentralized* management. A centralized management structure adheres to the example set in the military: decision-making power is concentrated under a single, central authority. The structure forms a hierarchical, multi-layered chain of command, and decisions tend to be made higher in the organization. Centralization necessarily implies that similar functions are undertaken in one location. Highly centralized organizations usually have a number of levels of management between top management and the line workers. Table 6.1 outlines some of the advantages and disadvantages of a centralized management structure.

In a decentralized organization, decision-making power is assigned to individual operating units. Decentralized organizations usually have fewer levels of management. The decentralization of a centralized company is accomplished first by removing some levels of management and then by delegating increased authority and responsibility to the remaining managers. In general, decentralization stems from the management philosophy that decisions should be made at the lowest level possible. To compare the advantages and disadvantages of decentralization, see Table 6.2.

In the final analysis, the best organizational structure synthesizes the benefits of both centralization and decentralization by combining the "advantages of having some form of purchasing decisions made at the location of the requirements, with the advantages of volume buying which come from handling common requirements on a centralized (aggregate basis)."[4] In a 1988 study, CAPS found that 59 percent of the responding organizations used some combination of a centralized–decentralized organization; 28 had adopted a centralized purchasing structure while only 13 percent were oriented in a decentralized fashion. The structure that a company actually chooses may depend on top management philosophy, the caliber of lower management, the nature of the business, or other factors.

SHOULD PURCHASING BE CENTRALIZED OR DECENTRALIZED?

Centralization versus *decentralization* in terms of supply management is concerned with the degree to which attempts are made to consolidate activities at higher levels of the organization. In particular, this is the extent to which the procurement activities are organized to coordinate requirements and combine orders. In the scheduling of work, Japanese factories often stress local responsibility, just as product development scheduling is

---

[4]Leenders, Michiel R. and Harold E. Fearon, *Purchasing and Materials Management,* Tenth Edition, Homewood, IL: Irwin, 1993, p. 49.

TABLE 6.1
Centralization: Advantages and Disadvantages

ADVANTAGES	DISADVANTAGES
1. Decisions are made by managers who have a broad view of the whole organization.	1. Decisions are made by managers who have a narrow focus.
2. Decision makers at higher levels are usually better educated and better trained than lower management.	2. Decision makers at the top rarely deal directly with the workers who must carry out the decision.
3. Eliminating duplication results in cost savings.	3. Long lines of communication cause long delays.
4. When functions such as purchasing are centralized, there is a greater chance for specialization which leads to increased skill levels.	4. Lower-level managers are often frustrated because they are removed from the decision-making process.
5. Decisions are more consistent throughout the organization.	5. Since more people are involved in communication, there is greater chance for error; additionally, personal bias & politics may enter the picture.

decentralized. The traditional Western approach requires centralized scheduling, plotting, and tracking.[5] Frequently, an effort to centralize all procurement,

---

[5]For additional information regarding centralization versus decentralization as it applies to purchasing, please see Corey, E. Raymond, *Procurement Management: Strategy, Organization, and Decision Making,* Boston: CBI Publishing, 1978; Kotler, Philip, *Marketing Management: Analysis, Planning, Implementation, and Control,* Sixth Edition, Englewood Cliffs: Prentice-Hall, 1988; Leenders, Michiel R. and David L. Blenkhorn, *Reverse Marketing: The New Buyer-Supplier Relationship,* New York: The Free Press, 1988; and Spekman, Robert E., "Strategic Supplier Selection: Understanding Long-Term Buyer Relationships," *Business Horizons,* Vol. 31, Number 4, July-August 1988.

TABLE 6.2
Decentralization: Advantages and Disadvantages

ADVANTAGES	DISADVANTAGES
1. Decisions are made faster.	1. There tends to be lack of information and coordination between departments.
2. The manager who has the most information about the situation makes the decision.	2. The cost is greater per manager because better-trained, better-educated, & higher-paid managers are required at lower levels.
3. Increased involvement in decision making creates higher morale and motivation among first line & middle managers.	3. Managers tend to have a narrow viewpoint and may be more interested in the success of their departments than of the company.
4. This structure provides good training for first line and middle managers.	4. Policies and procedures vary widely throughout the organization.

especially commodities, in order to leverage economies of scale in the marketplace and short-term schedules, typifies Western procurement. At the other end of the spectrum, integrated product teams make centralization versus decentralization decisions based on Total Systems Cost. Decisions are made based on who the major suppliers are across the board as opposed to those involved in individual projects. These efforts often drive process and product technology investments.

Total decentralization of purchasing would entirely eliminate the purchasing department. All departments would do their own buying. In a totally centralized purchasing department, everything is bought at one location by one purchasing department. Neither of these represents the usual case, though many small organizations have no purchasing department.

To what extent should purchasing activity be centralized at the organizational level? In practice, virtually every organization answers this question differently. Some centralize the activity almost completely, doing the buying for all sites at a central headquarters office. Others decentralize the function entirely, giving each site full authority to conduct all its purchasing activities. The majority of organizations, however, operate somewhere between these two extremes. Each approach offers significant benefits, which are discussed in the following sections.

Centralization is often used for one or all of the following reasons:

1. The centralized purchases are a very high percentage of product cost or budget.
2. The centralized items are used by the majority of the operating units.
3. Management feels the need for tight control of purchases and chooses to place the authority and responsibility into one central group, which often also assumes the responsibility for developing purchasing policy and control measures.

TABLE 6.3
Centralized Purchasing Advantages

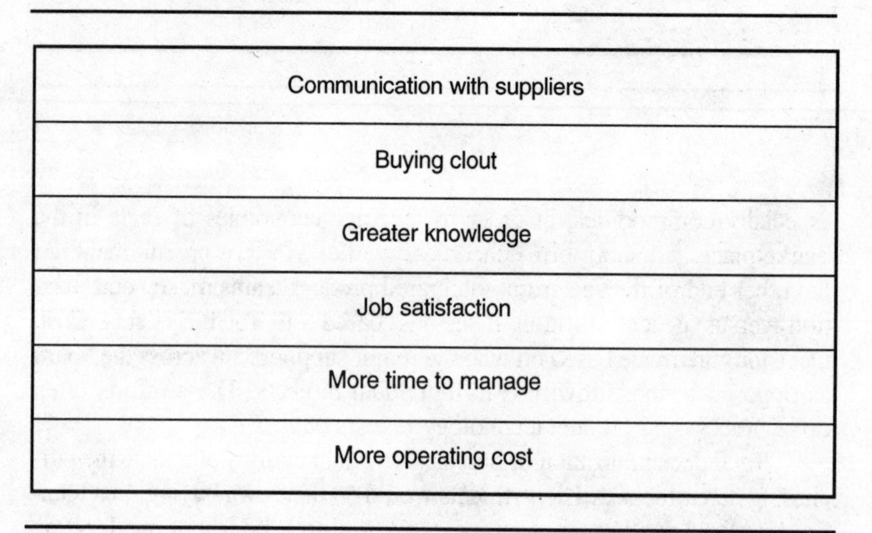

Centralized Purchasing Advantages
Communication with suppliers
Buying clout
Greater knowledge
Job satisfaction
More time to manage
More operating cost

Advantages of Centralized Purchasing

Centralization (which implies a multi-location company that does all or most of its buying in one location) has the advantages shown in Table 6.3 and described below.

Communication with suppliers—A central buying group becomes the spokesperson for the organization with its suppliers, thereby making it easier for both buyer and seller to make known their objectives and work together efficiently.

Buying clout—Centralized purchasing permits the consolidation of requirements for all operating units and usually leads to improved initial purchase prices and/or services. Sellers generally control costs more effectively when they sell to one department or buyer and know they are reaching the entire organization.

Greater knowledge—Centralization generally affords more specialization among the staff. This focus permits buyers to be more effective in obtaining value as they better understand what they buy.

Job satisfaction—The centralized department contributes to increased job satisfaction because buyers are able to make a more significant and measurable contribution to the organization.

More time to manage—Planning, organization, professional development, and personnel issues require time, and it is easier to allocate resources to these areas if the centralized department is adequately staffed.

Lower operating costs—Consolidation occurs in a centralized organization in terms of purchase orders, requisitions, telephone and written communications, sales interviews, and the coordination and scheduling of meetings. These combined efforts usually lead to lower operating costs.

Advantages of Decentralized Purchasing

As used herein, decentralization means centralizing purchasing at the local level; it does not imply total decentralization. An example of decentralized purchasing in a multi-plant organization with centralized corporate functional supervision at the plant level is shown in Figure 6-5.

FIGURE 6-5
Decentralized Purchasing (Multi-Plant Organization)

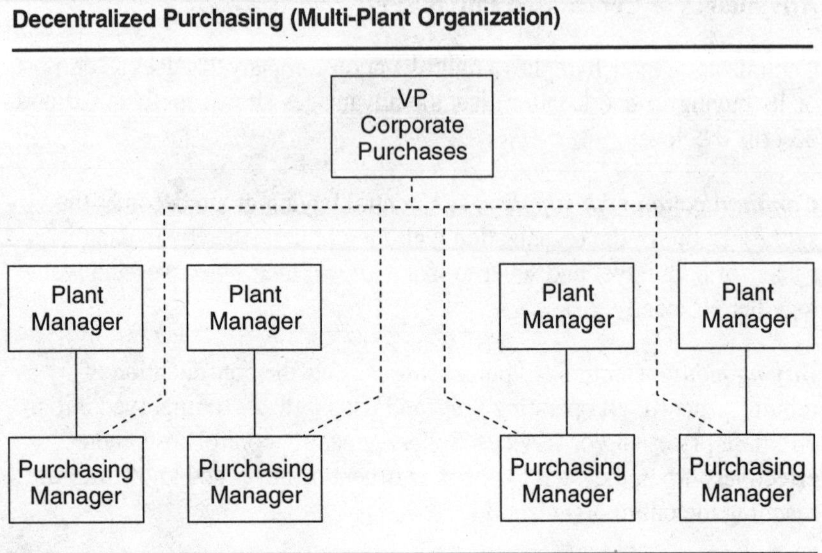

Decentralized purchasing has the following advantages:

Communication with internal customers. The closer buyers are to their internal customers, the better they will understand their needs and respond accordingly.

Broader responsibilities. The decentralized buyer must necessarily accept a wider range of responsibilities, quite often involving tasks that are not traditionally associated with purchasing (e.g., customer contact or operations liaison function).

More authority. The decentralized buyer often has more decision-making authority due to the lack of a reporting hierarchy.

Functional Authority of Decentralized Purchasing

Functional authority gives a manager the right to make decisions affecting employees and/or operations that are under the direct control of another manager. It is not uncommon for large corporations with multi-plant locations to decentralize corporate purchasing but to centralize local purchasing at each plant. The purchasing manager at the plant level usually reports to the plant manager. Often, but not always, these large corporations also have a corporate purchasing function. Sometimes this corporate entity is only one person, such as a corporate vice president of purchasing, while at other times, it is a small group of people (see Figure 6-4).

Many times, such as in the Maytag Corporation, the corporate purchasing group does not have direct authority over the local purchasing departments. They report to the head of the North American Appliances Group and are responsible for developing a competitive advantage by centralizing purchasing not only for Maytag appliances but for other product lines like Hoover. They therefore must either rely on their ability to persuade the local purchasing groups to get things changed or work through existing operations channels to the purchasing manager's or plant manager's supervisor to effect changes. To facilitate these interactions, Maytag recently developed a corporate purchasing council to address consolidation issues and centralized purchasing.

These corporate purchasing departments also provide other services such as: (1) establishing corporate policies and procedures; (2) auditing local plant purchasing departments to ensure that they are following corporate purchasing policies and procedures; (3) helping to hire, train, and promote purchasing people at the local level; (4) functioning as a resource for those seeking advice; and (5) acting as an advocate at the corporate level for local purchasing while similarly serving as the corporate voice at the local level.

Organizing for Global Sourcing

Because of the distances involved, methods for obtaining criteria on international suppliers are different than domestic means. An initial consideration is whether or not a company wants to contact foreign suppliers directly or to use intermediaries. The indirect approach usually has less risk but higher costs, and it is particularly useful for companies new to international procurement. This process involves import merchants, commission houses (handling consignments for foreign governments) and trading firms (import and export). For each, purchasing is essentially a domestic transaction. In any case, competition is encouraged, and such agents should be prequalified. The next step organizations often consider is the use of an import broker, where for a fee he or she acts as an intermediary for identifying suppliers in foreign locations. Shipping and alternate sources of supplies may be important.

Direct international buying involves completely coordinating the procurement activity; computers and EDI help but are not a panacea. The first step is to agree on contract terms. Legal recourse, quality control, conditions for order changes, cancellation clauses, and inspection and taxes are important considerations. On January 1, 1988, the United States signed the UN Convention on International Contracts, which is a useful convention (a type of uniform commercial code) in this arena, but it does not substitute for a viable contract between parties. Direct purchasing necessitates the consideration of the following:

- Legal ramifications (applicable federal and foreign regulations).
- Evaluation.
- Specification/pricing.
- Financing.
- Negotiation.
- Traffic.
- Entering through customs.
- Tariffs, quotas, licensing, patent and trademark laws.

Obtaining product information is extremely important and is often preceded by research, but frequently culminates in on-site visits (where company translators can be an asset). Planning oftentimes includes a study of customs and the culture of the countries involved. For example, it is useful to know that the Japanese take time to negotiate but are quick to implement.

One way of simplifying global sourcing is to use customs agents or indirect transactions (North American-based firms, import merchants, or foreign trading firms) in which the importer assumes most of the risks, often by accepting title for the goods in question. For a fee typically based on purchase value that may be as high as 25 percent, an import broker, sales agent, or import merchant will assist in locating suppliers and handling required paperwork. Trading companies usually handle a wide variety of products and services and typically target specific countries or regions. They are widely used by Japanese firms to move products around the world. Advantages of using a trading company include convenience, efficiency, lower costs based on volume, reduced lead times by maintaining inventory in North America, and higher quality, because the trading company assumes responsibility for incoming product.[6]

---

[6]Leenders and Fearon, p. 509.

On the other hand, direct purchasing involves a one-on-one contractual relationship which necessitates involvement in all aspects of the trading relationship (primarily for control). Advantages are lower cost materials, no major capital investment, and access to state-of-the-art technology or to stronger potential buyers. Limitations are longer supply lines and complex post-order services. Weights and measures and currency clarifications need to be made as well as an adaptation for cultural differences. Learning to speak the language is often a gesture of the commitment to the relationship. Source development is a critical facet of international procurement. Trade centers, the U.S. Department of Commerce (TSUS numbers), international directories such as Dun & Bradstreet, or international manufacturing directories (available in most libraries), international marketing departments or offshore banks can be valuable sources of information. Overseas trips should be team-oriented, and final reports should receive wide dissemination. The ultimate situation is to establish an overseas buying office.

Before buying offshore, it is necessary to ensure that global sourcing is consistent with organizational goals, that the long-term impact is understood and that contractual restrictions are examined. In addition, an understanding of countertrade, which is particularly attractive to cash-poor or centrally planned economies, is essential. Countertrade is basically a form of bartering and usually adopts one of five forms: barter/swaps, offset arrangements, counterpurchases, buyback/compensation, or switch trades.[7] Some of the concerns with countertrade are the transfer of technology advantage overseas, the loss of work, and the development of potential competition.

Because of the complexity of global sourcing issues, many organizations use international purchasing organizations (IPO). To effect its worldwide multi-billion-dollar sourcing operations, Thomson Consumer Electronics has 150 people in 15 offices around the globe. In addition to identifying and selecting sources, these IPOs often negotiate with the suppliers and, assuming agreements are reached, prepare the annual contract. Advantages of having foreign offices include:

- Better supplier relationships (facilitated by face-to-face cooperation).
- Lower procurement costs (no middlemen).

---

[7]Forker, Laura B., *Countertrade: Purchasing's Perceptions and Involvement,* Tempe, AZ: Center for Advanced Purchasing Studies/National Association of Purchasing Management, 1991, pp. 8-9.

- Enhanced buying opportunities.
- Improved quality control (on-site inspection and communication).
- Positive negotiating results (based on locally obtained information).
- Facilitated trade operations.

In terms of staffing requirements, the ultimate goal should be to use foreign nationals in these offices, but North American managers may be required initially. It recommends against using US-based foreign nationals because of the potential resentment. Local pride should be respected, and motivation should be by local custom (e.g., *shunto,* the Japanese Spring offensive to increase labor).

Before entering into an international relationship, it is necessary to: evaluate the opportunities (advantages in terms of price, quality, availability, and technology) that international procurement offers; examine the benefits; determine whether or not a separate purchasing department should be formed or whether it is necessary to open a foreign office (staffed by host country nationals or North American managers); and finally, consider the implications with regard to public relations. Reaping the benefits of an IPO is a long-term prospect that requires a significant amount of organizational resources to sustain. As a minimum, the firm should be buying more than $4 million per year in parts from the region, and anticipated savings should amount to at least 20 percent on the initial purchase price of these parts.[8] In an international context, understanding and respect are important, but an enhanced knowledge of international practices and customs is critical to successful international sourcing. The advantage of global sourcing is the forced development of an array of offshore relationships that might not be uncovered otherwise; but in the final analysis, landed costs in the context of total systems costs are the only reason to consider international procurement.[9]

---

[8]Carbone, James, "IPOs: Buyers' Windows on a Global Market," Electronics Purchasing, February 1990, p. 38.

[9]For additional information on global sourcing, please see the NAPM professional development films "International Procurement Part I: Deciphering Complexities" (Program Aids Library 94) and "International Procurement Part II: Developing the Skills" (Program Aids Library 95), produced in 1988 and 1989 respectively. These tapes offer a good introduction to global purchasing and discuss many of the concepts that have to considered when looking at foreign sourcing. They provide additional details on international procurement that may at last pique interest if not actually contribute to direct learning.

FIGURE 6-6
Departmental Organization

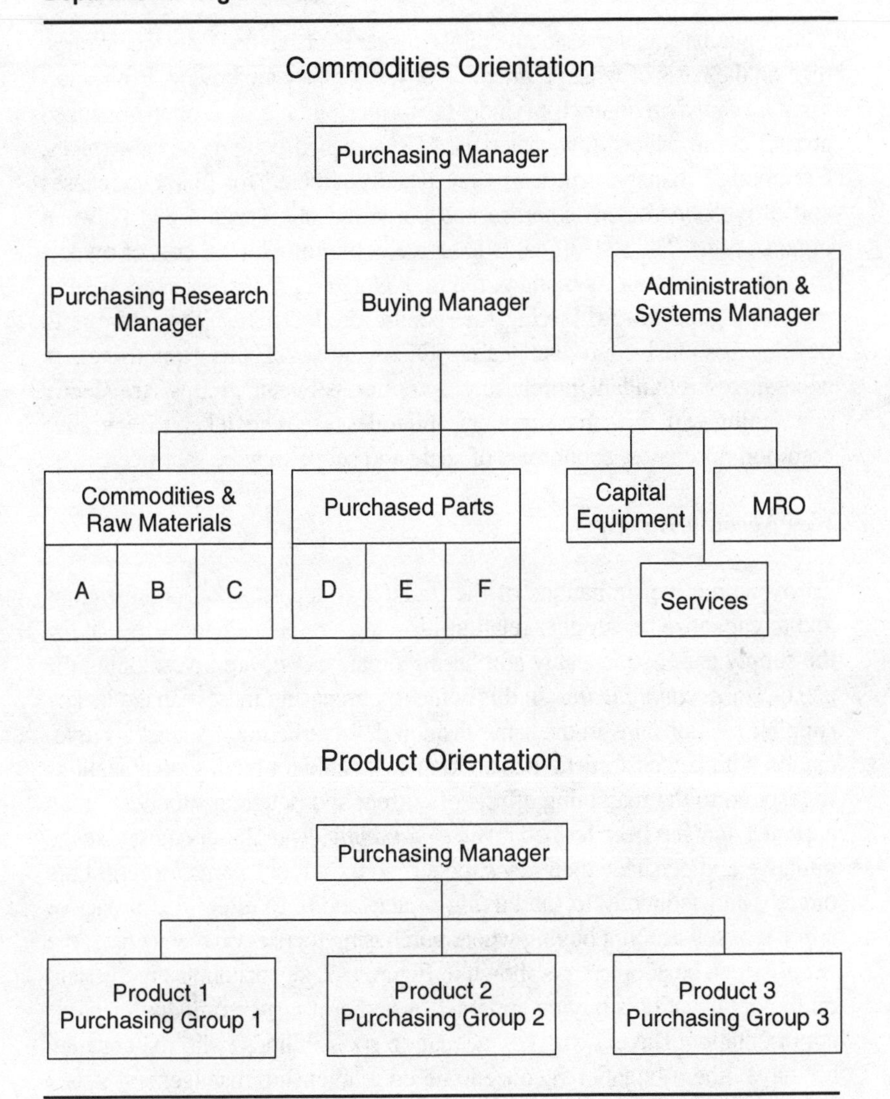

Departmental Organization

Within purchasing, the basic structural consideration has been whether to organize on the basis of what product line or service they are buying. Both structures are shown in Figure 6-6. Under the former, purchasing is often organized around commodities, raw materials, MRO, capital equipment, or services. Commodity management teams are usually adopted for major purchases and provide increased expertise and communication within and between organizations. The stated goal is to arrive at the lowest total cost of ownership for the commodity being purchased. Under a product orientation, buyers are aligned toward buying for specific product lines. The intent is to develop responsive product teams. Often, however, this responsiveness necessitates redundant purchasing activities between groups (frequently negotiating with the same supplier), and unless steps are taken to centralize common purchases, economies of scale and scope may be sacrificed.

Key Account Buying

To overcome the limitations of the traditional departmental organizations and to capitalize on supplier relationships and strategic alliances in making the supply team work, many purchasing organizations are investigating the use of *key account buying.* In this context, purchasing must manage its key supplier relationships in the same fashion that marketing manages its association with key customers. It is no longer possible for buying organizations to respond to the marketing efforts of current and potential suppliers. In an approach that has been termed *reverse marketing,*[10] purchasers must take the initiative and persuade their key suppliers so the supply team can contribute directly and indirectly to the buying organization. To effect this proactive approach, key account buying where purchasing focuses on the needs of the ***supplier*** can be adopted. As shown in Figure 6-7, key accounts are overlaid on the organization's buying approach, whether it is a commodity or product orientation. Buyers with key account responsibilities (who may or may not have other obligations) concentrate on relationship management issues rather than on individual purchases. Such approaches require teamwork and interpersonal skills and a commitment toward developing mutual interdependence between organizations.

---

[10]Leenders, Michiel R. and David L. Blenkhorn, *Reverse Marketing: The New Buyer-Supplier Relationship.* New York: The Free Press, 1988.

FIGURE 6-7
Key Account Buying Organization

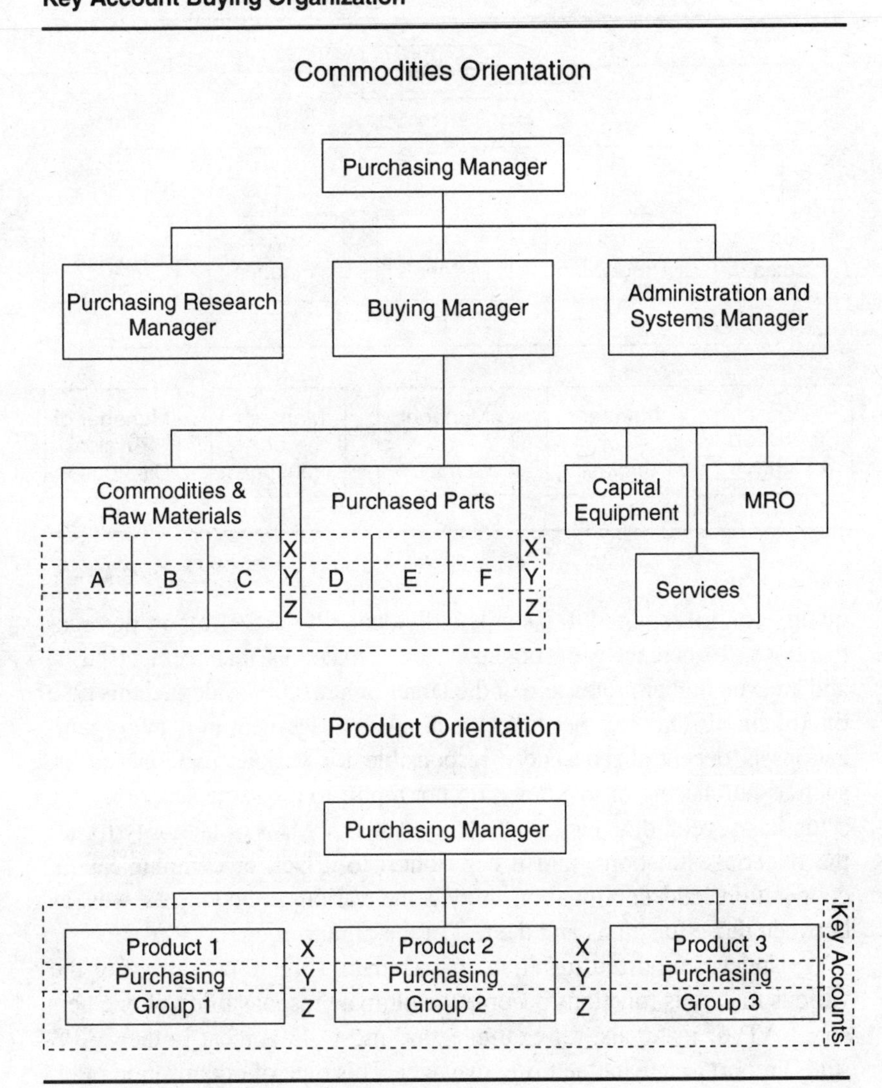

MATERIALS MANAGEMENT ORGANIZATION

A recent CAPS study defined a materials management organization as one in which at least *three* of the functions of purchasing, inventory, production scheduling and control, inbound traffic, warehousing and stores, and incoming

FIGURE 6-8
Centralized Materials Management

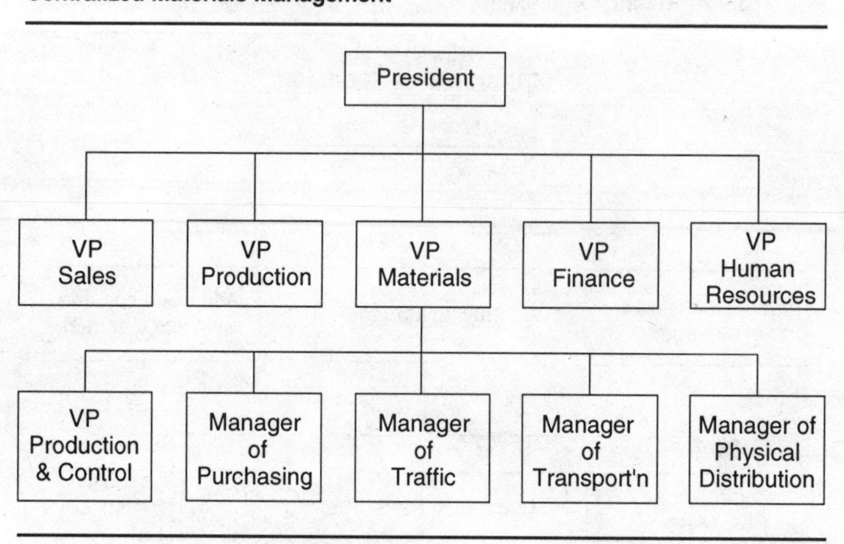

quality control reported to a single individual. Of the 291 reporting organizations, 70 percent were organized on a materials management basis, and an even higher proportion of the larger organizations adopted this type of structure.[11] On the other hand, materials management in many organizations is decentralized. Those responsible for various materials tasks, such as purchasing or inventory, do not report to the same superior. This setup has several drawbacks. Often, it results in a loss of authority for all the materials functions, and it contributes to a lack of communication, cooperation, and coordination among the various functions as well as between those functions and the rest of the organization.

Under centralized materials management (Figure 6-8), most of the various materials functions become the ultimate responsibility of one person, a VP of materials, for example, and most (43 percent in the CAPS study) report to a manufacturing manager. This type of organization often results in greater coordination within the materials division and in better materials service to the entire firm. It also gives the materials division

[11]Fearon, Harold E., *Purchasing Organizational Relationships,* Tempe, AZ: Center for Advanced Purchasing Studies/National Association of Purchasing Management, 1989, p. 17.

more clout in the operation of the company. It is important to reiterate that materials account for more than 50 cents of every sales dollar in most manufacturing firms.

Generally, there are five materials functions. *Production and inventory control* is responsible for scheduling production, planning raw material needs, requisitioning materials from purchasing and stores, and controlling finished goods inventory. *Purchasing* is responsible for buying the right quantities at the right time, place, and price. *Traffic* controls buying, scheduling, auditing, and billing of common and contract carriers. *Transportation* manages the firm's own transportation equipment and workers. *Physical distribution or warehousing* takes care of shipping, receiving, internal movement, and storage of raw materials and finished goods. Many companies are not large enough to justify a manager for each of these functions. In such cases, one manager can cover more than one function. For example, a manager can oversee purchasing and production control, or a materials VP can double as a functional manager.

Recently, however, integrated logistics management has also been the focus of significant attention. "Logistics refers to the art and science of obtaining and distributing materials and product"[12] and encompasses the flow and storage of items from raw materials to customer delivery. Although often used interchangeably with materials management, the core of integrated logistics is more on physical distribution and customer focus than on the up-front, incoming materials orientation typically associated with a materials management organization. This type of organization is especially useful for product lines where warehousing and distribution are paramount, and the emphasis is necessarily on the outbound strategy. In theory, the appeal of the integrated logistics concept is that

> it looks at the material flow process as a complete system, from initial need for materials to delivery of finished product or service to the customer. It attempts to provide the communication, coordination, and control needed to avoid the potential conflicts between the physical distribution and materials management functions.[13]

Practically, however, it will be necessary for purchasing managers and others in the supply team to overcome functional and organizational myopia. It is imperative to break down existing barriers, both real and imagined,

---

[12]APICS Dictionary, Sixth Edition, Falls Church, VA: American Production and Inventory Control Society, 1987.

[13]Leenders and Fearon, pp. 59-60.

to facilitate the evolution of requisite interfaces and interdependencies. Certainly from a system's perspective, the advantages of a fully interdependent supply team will exacerbate the use of integrated logistics organizations. Leading this progression toward greater integration within and between members of the supply team is the use of committees, task forces, work groups, and high-performance teams. The next section describes the growing impact these work units continue to have on the purchasing environment.

GROUP DYNAMICS

Group dynamics are having a fundamental impact on the way that purchasing organizes to make the supply team work. Groups often include committees, formal and informal work groups, task forces and teams. The ability to function and often to lead these teams can have a profound effect on purchasing's ability to contribute to organizational goals and objectives.

Committees

Committees allow more people to participate in the management process, thus increasing their motivation. Other reasons for having committees are (1) to improve coordination and communication between departments and/or divisions, (2) to promote group judgment, which is supposedly better than individual judgment, and (3) to provide checks and balances within the organization (as with a watchdog committee).

A committee is a group of individuals who have been assigned specific duties and responsibilities in addition to their normal work. As such, a committee usually represents a modification in the existing line, staff, and service organization and often is designed to handle certain standing activities (e.g., new product development, standardization, make-or-buy or purchasing research). Seldom does a committee replace a significant part of the overall organizational structure, although it legitimately deals with issues that cross functional lines. Purchasing frequently has an ongoing responsibility to serve on committees that cross organizational boundaries. In this capacity, purchasing professionals must represent the supply team.

Committee Membership

In serving on a committee, it is important to make remarks brief and to the point, to ensure that ideas are well formulated before speaking, to prepare for the meeting beforehand in order to make meaningful contributions, to take notes of important points made by associates, and—rather than arguing with other members—try to persuade them by asking questions that expose the weakness of their viewpoints. The role of a committee chairperson necessitates additional responsibilities:

1. Identify the task or tasks with which the committee is charged.
2. Establish an agenda. Mail the agenda to the other members in advance when possible.
3. At the beginning of the meeting, determine if anyone wants to change the agenda. The changes can be made informally by the chairperson or voted on by the membership of the committee if need be.
4. Once the meeting is under way, see that all the members follow the agenda.
5. Allow a reasonable amount of time for discussion by each member on every issue, but be careful not to let the discussion become too lengthy. Keeping the meeting moving along is one of the main responsibilities of the chairperson. (Remember that sometimes in the U.S. House and Senate, members are restricted to 30 seconds.)
6. After a reasonable amount of time for discussion, the issue should be put in the form of a motion or read if there is already a motion on the floor. Then the motion should be put to a vote.
7. Subsequently, the group's attention should be directed to the next item on the agenda, and the process should be repeated until all the issues have been adequately addressed.
8. The chairperson is also responsible for summarizing the key points of the committee meeting and for ensuring that adequate minutes are taken. Meeting minutes often serve an important role in record keeping.

Formal Work Groups

Formal work groups refer to the manager's relationship with individuals as members of a formal group (i.e., the department or functional sections within the department). Working groups are distinguished from teams in that there is *"no significant incremental performance need* or opportunity that would require it to become a team."[14] Typically, they exist to share information, best practices or perspectives and to make decisions that help individuals perform better. Often there is no genuine or desired common mission.

Informal Work Groups

Within every formal organization, experience demonstrates that one also finds the existence of one or more informal groups (also known as task forces or tiger teams). These groups typically are fairly small and are structured informally around specific interest patterns of the members. They may be educational, social, special interest, or sometimes pressure groups pushing for change. Whatever the case, such informal groups are an integral part of the departmental organization, and their attitudes and actions can either assist or deter the attainment of departmental objectives.

A wise purchasing manager attempts to exercise the potential influence of informal groups in a constructive manner. To do this, he or she must first recognize the existence of such a group and identify the informal group leader(s). Then, by practicing the concepts of open communications and group involvement in the decision-making process, the manager attempts to align the objectives of the informal group with the objectives of the department. This represents an extension of the participative management strategy employed in dealing with individuals and formal work groups within the department. Specific approaches that can be used include the following:

1. Through various individual, committee, and brainstorming techniques, solicit appropriate input on decisions that affect individuals and the informal group. To the extent possible, utilize this input in reaching genuine group-oriented decisions.

---

[14]Katzenbach, Jon R. and Douglas K. Smith, *The Wisdom of Teams: Creating the High-Performance Organization,* Boston: Harvard Business School Press, 1993, p. 91.

2. Create and develop a work climate and a reward system that encourage teamwork and cooperation.
3. Attempt to develop the degree of informal group cohesiveness that produces a positive influence on the activities of the formal work group.

The purchasing manager's goal in utilizing this integrative approach is to promote cooperation of the various groups—formal and informal—in daily activities that contribute to attainment of the department's overall objectives.

Teams

High performance teams consist of a small number of people with complementary skills who are equally committed to a common purpose, goals, and working approach for which they hold themselves mutually accountable. They are made up of members who are also deeply committed to one another's personal growth and success.[15]

Many organizations use teams to accomplish organizational objectives, including key elements of purchasing. "Teams outperform individuals acting alone or in larger organizational groupings, especially when performance requires multiple skills, judgments, and experiences....Nevertheless, most people overlook team opportunities for themselves."[16] Since making the supply team work necessarily implies the use of multiple skills, judgments, and experiences, purchasing offers significant opportunities for exploiting the role of teams. Some purchasing teams are made up solely of purchasing department staff, while some may include personnel from other parts of the organization, or even suppliers.

As an example, in Europe, sourcing plays an active role in Thomson Consumer Electronic's early-to-market core teams that are typically organized around new products. This tollgated product development process, which was adopted in 1992 for color television chassis, enables sourcing to become involved much earlier, typically during product concept, than had traditionally occurred. This core team concept, with the team leader at the center, is depicted in Figure 6-9. Team leaders have typically been

---

[15]Katzenbach and Smith, p. 92.

[16]Katzenbach and Smith, p. 9.

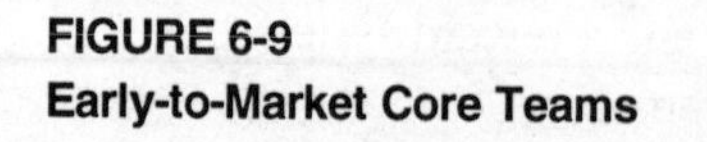

FIGURE 6-9
Early-to-Market Core Teams

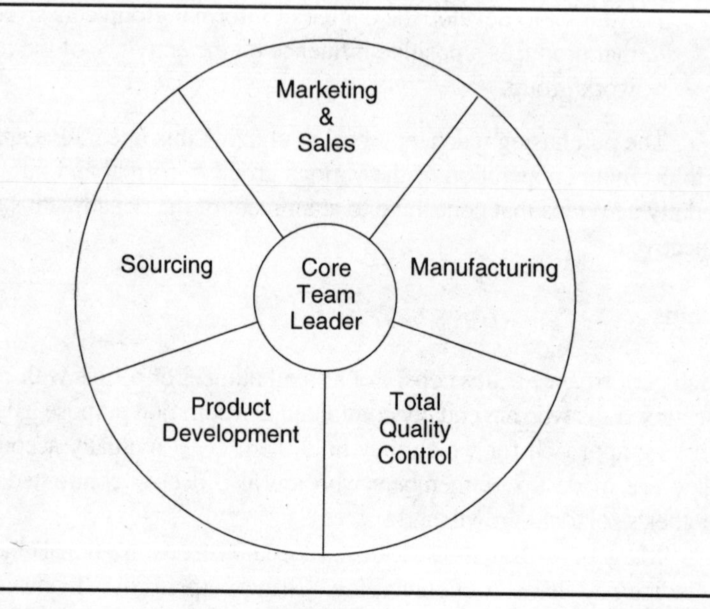

drawn from traditional product development organizations such as design or marketing, although in future finished goods procurement, such as a combination TV/VCR, Marc Latney, of Component Europe Sourcing, anticipates that sourcing may assume the role of core team leadership and will effectively manage the key account with their major supplier.

Figure 6-10 reflects the potential extent of teaming relationships typically encountered by purchasing with other members of the supply team. Regardless of its composition, teaming represents a collection of people who work or function together in varying degrees to achieve a common goal. Some recent insight regarding the application of teams is shown in Table 6.4.

The team-building process begins with identifying the need or reason for creating the team and the establishment of clear, compelling, and worthwhile goals or objectives, which can be accomplished only by a team. Team bonding, or the identification of and commitment to individual roles and

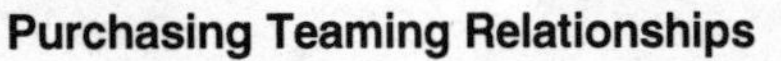

FIGURE 6-10
Purchasing Teaming Relationships

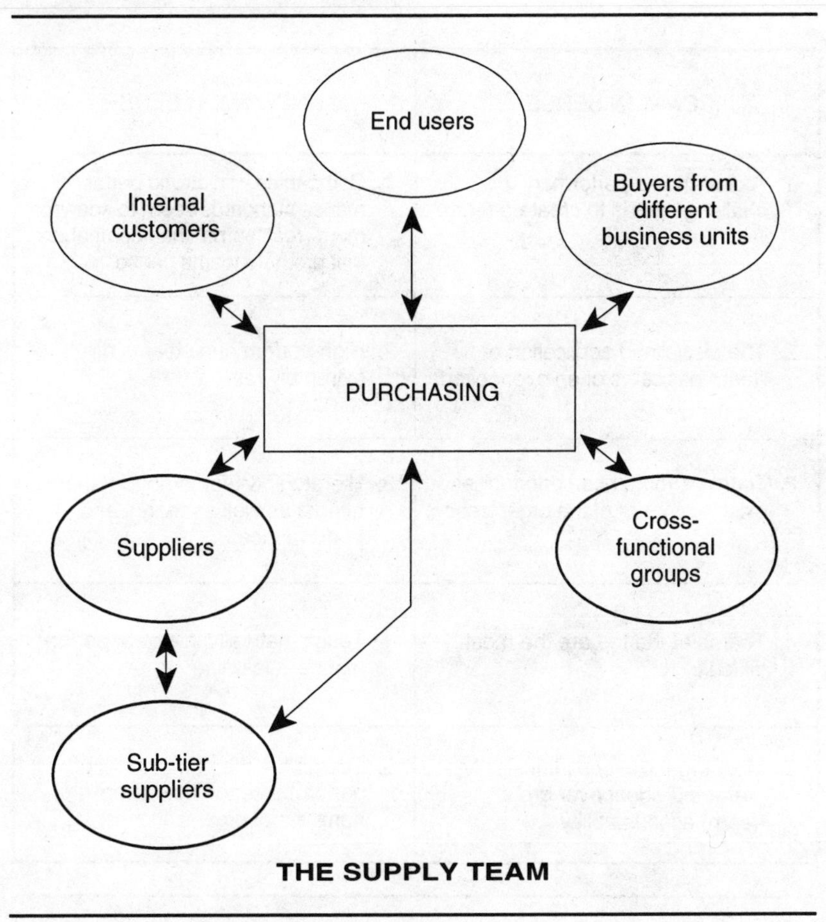

responsibilities, is essential. The team must have effective leadership to succeed, and there must be interdependence among team members to accomplish team goals. The team must be supported and empowered by management and be provided prompt, effective feedback on both individual and team performance.[17]

[17]"Teaming: Bringing it All Together," *NAPM Insights,* August 1991, pp. 12-13.

TABLE 6.4
Team Findings

COMMON SENSE	UNCOMMON SENSE
1. A demanding performance challenge tends to create a team.	1. Companies with strong performance standards seem to spawn more "real teams" than companies that promote teams per se.
2. The disciplined application of "team basics" is often overlooked.	2. High-performance teams are extremely rare.
3. Team performance opportunities exist in all parts of the organization.	3. Hierarchy & teams go together almost as well as teams and performance.
4. Teams at the top are the most difficult.	4. Teams naturally integrate performance & learning.
5. Most organizations intrinsically prefer individual over group (team) accountability.	5. Teams are the primary unit of performance for increasing numbers of organizations.

Source: Adapted from Jon R. Katzenach & Douglas K. Smith, *The Wisdom of Teams: Creating the High-Performance Organization,* Boston: Harvard Business School Press.

In the context of making the supply team work, purchasing professionals must also effect collaborative relationships with key customers and suppliers. The following example provides an illustration of the emerging role of suppliers in teams.

> Suppliers are also a critical constituency at Motorola. At GEG [Government Electronics Group], for example, materials and supplies account for more than half of the cost of doing business and cut to the heart of its ability to produce

> hundreds of different kinds of electronic systems and equipment for NASA, the U.S. Department of Defense, and other governmental and commercial customers. In fact, the Connectors Team actually grew out of GEG's effort to partner ore effectively with suppliers. In 1989, GEG's leadership team decided to shift supply management from a decentralized, functional organization that depended on the expertise and performance of individuals to a centralized, process-oriented organization that depended primarily on teams.[18]

To capitalize on the inherent advantages that teams bring to the organization, purchasing managers should initiate open-ended discussions about the performance and purpose that can lead to team development. It is often necessary to reevaluate the group goals; they should be clear, specific, measurable, and performance-focused. It is important to reconcile the goals with specific work team products that will produce desired results. In addition, managers need to focus on skills and attitudes rather than styles or personalities. One should never be complacent about the skills of the group; deficiencies must be confronted and rectified immediately. Achieving goals rather than viability of the team should be paramount.

In summary, groups serve a critical role in achieving interdependency. It is wise for purchasing professionals to become actively involved in groups that are addressing meaningful issues affecting the organization. Often these cross-functional opportunities and the inherent flexibility they require allow managers to leverage their individual skills and the capabilities of their organizations in value-adding efforts, with long-term implications.

THE PRINCIPLES OF ORGANIZATION

This section examines the principles of organization as they are used to make an organization function well. As an illustration, a highway system is analogous. In this context, the organizational structure is equivalent to the streets, roads, and expressways of the system, while organization principles are the road signs, traffic laws, and stoplights that make it possible to travel the system. In essence, the principles of organization are the rules of the organization road.

---

[18]Katzenach, Jon R. & Douglas K. Smith, *The Wisdom of Teams: Creating the High-Performance Organization*, Boston: Harvard Business School Press, p. 184.

Importance of Good Staffing

The basic principle of staffing is an appropriate starting point: *An individual who is appropriately trained or trainable, and properly motivated, must be secured for each job.* The success of any organization depends more on having the right people than on anything else. This is applicable whether the organization is a church, club, purchasing department, dry cleaner, or factory. A manager needs to take great care in the selecting, training, treating, and motivating of the people in (and often outside) his or her organization.

Subsequent chapters specifically address appropriate methods of selecting and training individuals and organizations. At this point, it is important to acknowledge that without good staffing, it is not possible to have a truly effective organization.

The continued life of the organization comes first. Therefore, each employee must have the ability to fill a place within the organization, rather than the organization changing to fit the employee's ability. More important than any individual benefits the organization might be able to provide for its employees is the continued existence of the organization. If the organization fails, no one benefits. The company's chief obligation to employees, customers, suppliers, and other stockholders is to stay alive and profitable.

From this perspective, it is essential that each new employee should strengthen the organization; an organization is only as strong as its weakest link. Major qualifications for each position must be met, but flexibility in how duties are performed should be expected as a result of an individual's personality, training, interest, and motivation. Because the positions within an organization are interdependent, an employee with inadequate qualifications causes degraded performance. Realignment of associated positions then becomes necessary. The realignment of job duties is sometimes desirable and necessary. However, one cannot shake up the organization every time an employee is replaced.

The Employee's Responsibility

By agreeing to be hired, employees obligate themselves to perform their job assignments properly under the direction of their superior. Correspondingly, the employer agrees to pay for these services. Either party has the right to terminate the agreement according to the terms of the contract. Some new managers do not understand the principles of responsibility clearly and are embarrassed about asking employees to work. It is not a crime to demand performance. Employees know that they are expected to work for their pay, and they will have more respect for managers who expect them to earn their money.

Managerial Authority

Managers have the right to perform their assigned duties and to command their subordinates. There are two kinds of authority. The first is *formal authority,* which is given to a manager by those who hold control of the organization. In other words, authority and responsibility are passed through the organization from superior to subordinate, and this establishes the reporting relationship. Reporting relationships are known as the *chain of command.*

The second kind of authority is *earned authority.* This is the power given to an individual in a group by the members of that group. It is hidden authority. It does not appear on organization charts, nor are there job titles that describe it. One of the biggest mistakes that some managers make is to assume that because they have been formally named to a position, employees will automatically follow all their orders. If they persist in this mistaken notion, they will usually be less effective.

To avoid this problem, it is important to establish earned authority as soon as possible. Earned authority does not come easily or quickly, and it must be pursued continuously. Earned authority is the natural result of employees accepting leadership because they have confidence in someone's ability and feel that individuals are sensitive to their needs.

The Principles of Delegation

Delegation can be defined as the assignment of responsibility and authority to a person who is acting as one's representative. There are five principles that make delegation more effective:

1. The principle of span control
2. The principle of fixed responsibility
3. The principle of equality of authority and responsibility
4. The principle of lower-level assignment
5. The principle of reporting exceptions

The concept of span of control suggests that there exists an ideal number of employees that a manager can effectively manage. All organizations would remain small if it were not for delegation, simply because there is a practical limit to the number of activities and people one person can effectively manage. We call this limit the *span of control.* If an organization is to grow, those in charge must be willing to delegate some of their responsibilities and authority to subordinates.

There is no way of calculating the ideal number of employees who should report to each manager. Instead, it is necessary to apply common sense and logic to arrive at a workable number. The main factors affecting span of control are as follows:

1. ***Ability of the Manager.*** Some managers are more capable than others and can, therefore, handle a larger span of control. In planning an organization, the span of control should be based on a manager of average ability.
2. ***Ability of the employees.*** Workers with less ability and training require more supervision than employees who are better qualified.
3. ***Type of work.*** If the work is machine-paced or repetitive, the employee requires less supervision. In other words, a supervisor can direct more employees if they are working on an assembly-line operation than if they are working in a purchasing department.
4. ***Geographic location.*** A purchasing manager who has 25 employees all located in one room may be able to supervise them effectively. But a sales manager who has 25 salespeople located in 25 different states would find direct supervision impossible.
5. ***Employee motivation.*** Employees who get little or no satisfaction from their jobs are poorly motivated and require close supervision. Highly motivated employees such as those at the Lincoln Electric Company, which is well known for its profit-sharing plan, need little supervision (span of control is 100 to 1).
6. ***Importance of the work.*** The person with the authority to decide whether to buy $100 million worth of equipment is in a stronger position to help or hurt the company than someone who has the authority to spend $1,000. Therefore, the custom is to have fewer people reporting to the president of an organization than to the first-line supervisor, because more attention should be given to those who make critical decisions.

Changes in span of control typically require top-level management intervention. The best way to demonstrate a need for the change is to construct a proposal based on any or all of the six factors listed above.

To succeed, the proposal for a change should address tangible and intangible benefits. Often, it is necessary to be patient yet persistent. Such changes are rarely taken lightly and should be based on facts rather than opinions.

Task Responsibility

Delegating responsibility to a subordinate does not relieve the delegator of the responsibility. Managers cannot delegate their responsibilities. The responsibilities remain fixed (or attached) to the managers to whom they were originally assigned. The process of delegation is a process of sharing responsibility with subordinates. The only way managers can be released from responsibility is if a superior takes it away from them.

> The purchasing manager asks a buyer to take an inventory and have the results by 8:00 A.M. Monday. The buyer delegates the responsibility for taking the inventory to an assistant, who forgets to take the inventory. Who is responsible? They both are. The buyer was not relieved of the responsibility by assigning the task to a subordinate, and therefore must share the responsibility with the assistant.

The level of authority delegated must equal the amount of responsibility delegated. In delegating, the most frequent mistake is failing to impart enough authority to a subordinate to ensure the proper fulfillment of the responsibility. As an example, assume that the mayor of a city delegates the responsibility for effective waste collection and street maintenance to the city service director, but for political reasons retains the right to hire and fire all employees. Assume further that the mayor has, on a number of occasions, refused to fire employees who are completely unqualified even though the service director has strongly urged that this be done. Should the service director be held responsible for the effective operation of the service department? No, because the mayor has failed to delegate enough authority to permit the service director to effectively discharge his or her responsibility. Will the service director be held responsible for nonperformance of the department? Probably. Although this is hardly fair, it happens time and again, not only in government but also in business and other organizations. When confronted with a situation in which responsibility is substantially greater than authority, an individual should be compelled to do everything in his or her power to petition the superior for adequate authority. If the necessary authority is not forthcoming, strong consideration should be given to finding a new job. Without authority equal to responsibility, failure is imminent.

The failure to delegate enough authority usually arises for one of three reasons:

1. Political reasons, as in the case above.
2. A lack of faith in one's fellow human being—the feeling that to do the job right, the individual alone must do it.
3. Desire for power—the reluctance of managers to give up any authority because to do so would make them feel less important.

Work should be assigned to the lowest organization level that can properly do the work. Economists call this the principle of absolute advantage. In other words, a $40,000-a-year buyer should not sweep and dust the office every day when a custodian earning $5.00 an hour can do the work. The principle of lower-level assignment seems logical, simply a matter of common sense, yet most people may know of a college president who sorts the faculty mail each day, deans who type their own letters when they have secretaries sitting outside their doors, and professors who grade true-false exams that could be graded by a student assistant for $4.75 an hour. In short, many people often violate this principle, and it is probably the biggest single contributor to under-used labor in America today.

This principle is usually violated for one of two reasons. The person either does not know the principle or succumbs to the natural desire to retreat to the familiar. The fact that people retreat to the familiar means that they are more comfortable doing things they know how to do because of past experience. This often happens when a person is given a promotion. A buyer who is made purchasing manager frequently wants to continue buying rather than managing the work of the department. A rate clerk who becomes traffic manager may continue to work on rates rather than finding ways to run the traffic department better. Professors often staple their own exams instead of finding ways to improve their lectures.

It is human nature to avoid creative work, because creative work is hard work. Instead, most people tend to seek routine work because it is easy. It is important to overcome this natural desire to retreat to the familiar by examining the contributions of each individual. It is important to delegate work to the lowest level possible.

All deviations from plan or expectations should be reported immediately. The idea is that an employee should report only the unusual things to a superior. A sailor on watch would be expected to report to the captain if the ship were sinking, or, in the case of the *Titanic,* that an iceberg was approaching. But the sailor need not report every half hour that the ship is afloat unless there is some question about the ship's seaworthiness.

Too often managers take up their supervisors' valuable time telling them that everything is okay. Just as infrequently do supervisors discourage this behavior. In brief, it is important to note all significant exceptions—good or bad. Under normal conditions it is not necessary to report the fact that everything is going according to plan.

Effective Delegation

To delegate effectively it is important to reexamine the five principles of delegation as they apply to each organization. As noted previously, span of control is a good starting point. Are subordinates being asked to supervise too many people? It is important to remember that the responsibility delegated to subordinates remains a manager's responsibility. It is necessary to work *with* subordinates to help them meet these responsibilities, but not to *do* their work for them. The key point is to *help* them develop their own capabilities. And it is also important to remember that if they look good, the supervisor looks good. It is not necessary to take credit for their accomplishments.

It is also essential to be honest. Has an equal amount of authority and responsibility been delegated? If not, the subordinates will probably be unhappy and inefficient. Managers must also question whether they are retreating to the familiar. Has the work been delegated to the lowest possible level? It is important to remember that the role of a manager is to get things done through others.

Subordinates and managers alike should be encouraged to report exceptions immediately and to reduce the number of reports of things going according to plan. Some additional pointers to keep in mind when delegating work are as follows:

1. *Be sure that subordinates know precisely what results are expected.* By explaining the job to them and then asking them to explain it or demonstrate it can be an effective method for ensuring that the task is understood. It is often not enough to simply ask them if they understand. They will almost always say yes either way. Either they mistakenly think they understand, or they are afraid to admit that they do not.
2. *Set a target date for completion.* If there is a complicated project that must be done over a period of months, work out

a timetable with the subordinate that shows when each phase is to be completed.

3. *Take the time to train subordinates to take on new responsibilities.* Often managers do not delegate additional responsibilities to a subordinate because the subordinate has not been trained. Managers have the attitude that it is easier to do jobs themselves than to train someone else, and it probably is for that one time. However, if a job comes up often, the manager would be better off taking the time to train someone else.
4. *Guard against unofficial delegation.* Delegation is often considered to be downward to a lower level. This is official delegation. But there are two types of unofficial delegation that should be avoided. The first is horizontal and occurs when someone on the same organizational level tries to delegate work. The second is upward and usually happens when subordinates delegate work up the chain of command.

When a subordinate comes to you with a problem (thus passing it up the chain of command), it is easy for you to say, "Don't worry about it. I'll find out the solution and get back to you." It is often unwise to accept a subordinate's problem; the supervisor can and should help if needed, but should resist the temptation to do the work. Some managers get a sense of satisfaction out of solving other people's problems, but a manager's job is to manage, not to put out brush fires. A manager's job is to provide the necessary fire fighting equipment and to teach people to extinguish their own fires.

Asking the subordinate what his or her own ideas are (usually the answer is hidden among these ideas) allows a supervisor to help without directly solving the problem. If more information is needed, helping the subordinate find out where to get it is also useful. On the other hand, sometimes a problem indicates that the department is not as well managed as it should be. In that case, analyzing the root cause of the problem and correcting it (often after helping to solve the immediate problem) can lead to continuous improvement over the long term.

How an Employee Knows What to Do

Each employee's duties must be clearly distinguishable from all other workers' duties, and the employee's duties must be known and understood by the employee. All too often, supervisors give instructions that are too general. Later, when they find out that everything they expected was not done, they, blame the employee for their own poor performance. Employees must know exactly what is expected of them. It is not enough to say, "buy this product," because the employee may seek three bids when the manager simply wants him or her to negotiate with the existing source.

It is an important duty of management to pinpoint responsibility. This is why people find little slips of paper in new clothes. The company is trying to ensure that worker #5 is doing a good job of producing a quality garment. Having a goal or task in mind is motivational, particularly when the individual is involved in establishing the goal. Taylor argued, "There is no question that the average individual accomplishes the most when he either gives himself—or someone else assigns him—a definite task; namely, a given amount of work which he must do within a given time."[19]

How an Employee Knows Whose Orders to Follow

No one in the organization should have more than one supervisor. This idea is that no one can serve two masters. Having more than one superior confuses accountability. This principle is most often violated with regard to the department secretary. Officially the secretary reports to the department head, but unofficially everyone in the office assigns him or her work. It is hard for the secretary to know whose work comes first. If dual accountability cannot be avoided, every attempt should be made to define functional authority clearly.

One of the most annoying things that can happen as a manager is to have the next level supervisor continually giving instructions to subordinates. Not only does it confuse accountability, but it also destroys morale and causes subordinates to lose confidence.

Organizational Policies

Purchasing must be sensitive to overall organizational policy. A complete understanding of the organization's mission and goals is necessary. This includes implementation of policy not always found in the purchasing manual, but in the overall organization policy, or policy from functional areas such as safety, operations, quality assurance, and human resources.

---

[19]Taylor, Frederick W. *Scientific Management,* New York: Harper & Row, 1947, p. 69.

KEY POINTS

1. Organizing begins by establishing a structure of relationships that facilitate the accomplishment of organizational goals and objectives. It is generally necessary only when a change of goals or objectives has occurred or a new plan is put into effect. Managers must determine what people and resources are available, as well as what the department's requirements and limitations are.
2. When organizing a department to perform a specific program or plan, a manager must first interpret the corporate goals as they apply to the department's functions. The manager determines what activities are needed and when, delegates these activities to department staff, and assigns responsibility for these tasks.
3. When organizing a purchasing department, a manager must evaluate the staff, assess their strengths and weaknesses, and determine what training or resources the staff needs to carry out their duties.
4. The structures of most organizations can be characterized as line, staff, or service. The line performs the primary functions of the organization while the staff supports the line. Service provides an indirect contribution to the primary line operations as well. A company must use open and supportive communications to ensure cooperation between line, staff, and service.
5. Within an organizational structure, line authority refers to the right and responsibility to command subordinates. Functional authority is the right to make decisions affecting operations of departments under direct control of another manager.
6. In a centralized management structure, decisions are made at one point, usually higher in the corporate hierarchy. Decentralized organizations usually have fewer levels of management, with the lower levels or operating units having more decision-making authority. An optimal corporate structure contains aspects of both centralized and decentralized organizations.
7. Benefits of a centralized purchasing department include cost savings, better buying clout through economies of scale, and better communications with suppliers.
8. Materials management usually consists of five operations: production and inventory control; purchasing; traffic control; transportation; and distribution. Integrated logistics is focusedmore toward physical distribution and meeting end customer needs.

9. Although their effectiveness is often questioned, the use of committees is increasing since they facilitate interdepartmental communications and promote group decision-making. Key points to remember when serving on a committee are: keep points brief; be prepared and organized; take notes; and ask questions to investigate ideas.
10. Managers must be aware of the various group dynamics within an organization. Informal work groups provide an opportunity for the manager to align the objectives of the group members with those of the organization.
11. Effective teams typically incorporate a variety of skill sets and focus them on achieving a common goal. A manager can help develop a team and keep it focused through open communications and guidance.
12. When developing an effective purchasing organization, the principles of organization should be followed. The principle of staffing ensures that trained, motivated employees are obtained for the various job positions. New employees should strengthen or add benefit to organizations, and they must agree to perform under the direction of their superior.
13. A manager is given functional authority by the manager's supervisors, thus developing a hierarchical reporting relationship called a chain of command. Earned authority, which a manager must continuously develop with his or her subordinates, enables the manager to function more effectively.
14. When determining the ideal number of employees who should report to a manager, consider both the manager's and employee's abilities, along with the nature, importance, and location of the work.
15. Responsibilities are never transferred to subordinates but are shared between them and their manager. When delegating responsibility, a manager must delegate an equal amount of authority to ensure that tasks or duties are performable.
16. Routine or repetitive work not requiring creativity should be delegated to the lowest possible level. Employees should limit reporting to relevant or important issues.
17. To effectively delegate work, a manager must share responsibility with subordinates, communicate the expected results, set a time line for completion, train the employees, and help them develop their own solutions to problems that arise.

18. A manager must provide a clear understanding to employees of what their duties and responsibilities are. The employee must know to whom they are responsible and who has authority over them.

REFERENCES AND RECOMMENDED READINGS

Katzenbach, Jon R. and Douglas K. Smith, *The Wisdom of Teams: Creating the High-Performance Organization,* Boston: Harvard Business School Press, 1993.

Scholtes, Peter R. *The Team Handbook: How to Use Teams to Improve Quality,* Madison, WI: Joiner Associates, 1988.

CHAPTER 7

STAFFING

STAFFING

Staffing purchasing is a critical facet of making the supply chain work. Recognizing that management implies getting work done through others necessarily places a significant emphasis on finding, selecting, hiring, and training new employees and developing and upgrading the skills of existing employees. Effective purchasing staffing can prevent many of the common employee problems. This process is complicated by the fact that new nontraditional skills will be required by purchasing professionals in order to add value within their organizations. A department's efficiency and, more importantly, its effectiveness depend on having the right people performing the right tasks. Purchasing managers must effectively forecast their personnel needs in terms of quantity and quality and must aggressively pursue purchasing professionals who will contribute to overall supply chain management goals and objectives.

It is essential that companies put their people first, and according to corporate leaders like Hal Rosenbluth, CEO of Rosenbluth International (the third largest travel management company in the United States with revenues in 1992 in excess of $1.5 billion), even before their customers. "We're not saying choose your people over your customers. We're saying focus on your people *because* of your customers. That way everyone wins."[1] What this means to purchasing is that staffing considerations must be contemplated in light of forging an organizational culture that contributes to the well-being of your people. Rosenbluth's hierarchy of concerns is: people, service,

[1]Rosenbluth, Hal F. and Diane McFerrin Peters, *The Customer Always Comes Second,* New York: William Morrow, 1992, p. 25.

TABLE 7.1
Cultivating Happiness in the Workplace

Cultivating Happiness in the Workplace
Happiness in the workplace is key to providing superior service.
People are companies' one true competitive measure.
Measuring happiness in the workplace is essential.
Make contact with your people at every level.
Make your company a lifestyle—not just a place to work.
Recheck the ratio of financial to humanistic pillars in your company's foundation.

Source: Hal F. Rosenbluth and Diane McFerrin Peters, *The Customer Always Comes Second,* New York: William Morrow, 1992, p. 39.

profits—in that order. Companies focus on their people who subsequently focus on serving the clients, with profits as the end result. Their keys to achieving happiness in the workplace are shown in Table 7.1. Making the supply team work, quite simply, starts with people.

Job Requirements

Skills Requirements

A departmental organization plan should serve as the blueprint from which all employee selection activities are developed. A sample employee selection process is depicted in Figure 7-1. To determine the most effective organizational and authority relationships within the department, each job must be carefully analyzed and described. This type of job analysis pinpoints the specific duties and responsibilities entailed, any unusual working conditions involved, and specific qualifications and characteristics required of the person holding the job. The latter determination focuses on the skills, abilities, knowledge, training, experience, and personal qualities necessary to do the job satisfactorily. Most organizations condense these findings and include them, for each job, in a written job description.

FIGURE 7-1
Employee Selection Process

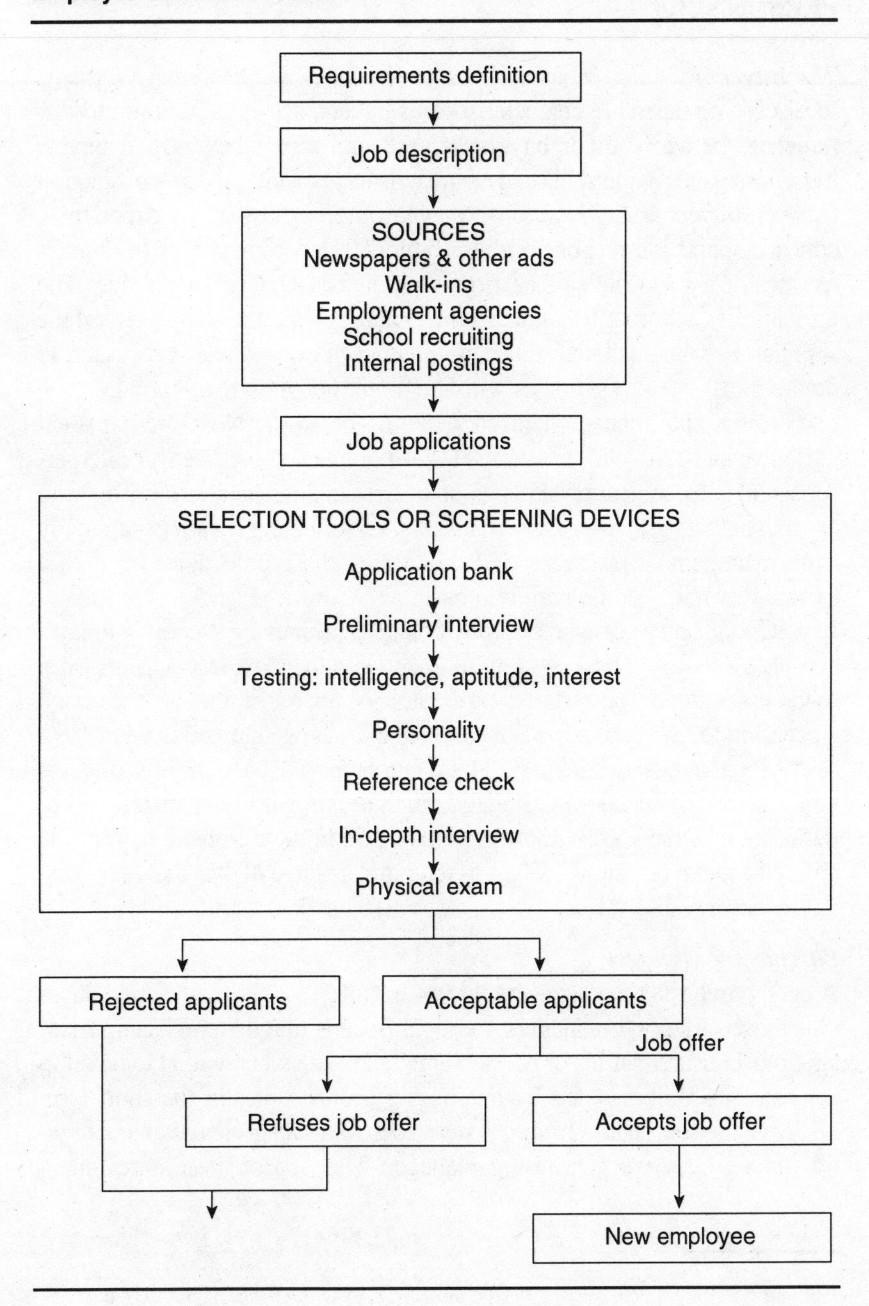

GENERAL CHARACTERISTICS FOR PURCHASING PERSONNEL

The Buyer

In discussing desirable characteristics of buyers, it is important not to differentiate between junior buyers, buyers, and senior buyers. All buyers must possess the same skills and characteristics; however, the key distinction between buyers at any level of the organization is a matter of experience, education, and job responsibilities. Table 7.2 lists four groups of characteristics of a good buyer. The first is product or service knowledge. The term implies a thorough technical understanding from the perspective of the supplier *and* the customer about the product or service one is expected to buy (depth). The second deals with the professional body of knowledge in purchasing and management in general (breadth). Next are personal attributes and core values. And the last set of characteristics cover interpersonal and communication skills. In an ever-changing business climate, buyers must act as global process managers, coordinating a variety of supply team management processes and monitoring these procedures on both a commodity and specific part/item basis.

Overall, interpersonal skills are critical for achieving success in today's complex interactive business environment, and their mastery appears even more essential as purchasing professionals approach the next century. According to the Purchasing Education and Training study released in 1993 by CAPS, the most desirable skill for a purchaser will be in the area of interpersonal communication. Currently, purchasers regard interpersonal communication as the second most important skill in their profession, after the ability to make decisions.[2] The top ten purchasing skills for the year 2000 are shown in Table 7.3.

Purchasing Manager

A good purchasing manager should possess all the qualities of a good buyer plus those of a good manager. A critical issue is that the purchasing manager must ensure that his or her personnel have the wherewithal to develop the requisite skills needed by purchasing professionals in the short term and over the long haul. Table 7.4 describes four categories of characteristics for an effective purchasing manager. The first of these is technical

---

[2]Ballew, Arlette C., *"The Need for Interpersonal Skills," NAPM Insights,* August 1994, p. 18.

TABLE 7.2
Characteristics of a Buyer

Product knowledge	Principles of purchasing & management	Personal attributes	Interpersonal skills
• Knows major facets of product or materials	• Knows purchasing's role in the organization	• Integrity	• Can work well in teams
• Knows the market price, etc.	• Understands quality theory & application	• Good mind for details	• Has good communications skills: – good listener – persuasive speaker – good report and letter writer
• Knows major sources	• Understands pricing theory, policies, & practices	• Likes to do research	
• Is familiar with quality requirements and problems	• Is a good negotiator	• Careful and deliberate decision maker	• Is willing and able to deal effectively with people who hold opposing views
• Understands external customer requirements and their impact on internal customers and suppliers	• Has a workable knowledge of inventory control & management	• Ability to tolerate conflicts and ambiguity	
	• Understands and can apply the right time concept	• Takes appropriate risks	
	• Is able to successfully carry out value analysis projects	• Has high self-esteem	
	• Knows purchasing role in capital equipment buying	• Takes initiative	
	• Is able to apply make-or-buy theory		
	• Is able to apply supply chain management concepts		
	• Knows marketing, accounting, MIS, operations, organizational behavior and financial management theory & practice		

TABLE 7.3
Purchasing Skills in the Year 2000

Interpersonal communication
Customer focus
Making decisions
Negotiating
Analyzing
Managing change
Conflict resolution
Problem solving
Influencing & persuading
Computer literacy

knowledge, which can be defined as basic knowledge about both the department itself and operations in general. Next is analytical ability, which is the ability to think in the abstract, solve problems, and make decisions. The third set of characteristics are interpersonal and communications skills. While in some ways these abilities closely resemble those needed by a good buyer, they are far more encompassing because of additional information sharing requirements placed on the purchasing manager. An effective purchasing manager is approachable and is a good listener; he or she communicates information in a positive fashion so that it is readily received and understood and can be acted upon quickly. The final group of characteristics

TABLE 7.4
Characteristics of a Purchasing Manager

TECHNICAL KNOWLEDGE	ANALYTICAL ABILITY	INTERPERSONAL SKILLS	MANAGERIAL SKILLS
• Product knowledge—has a good understanding but not necessarily an expert in every product bought by the department • Has mastered the principles of purchasing • Is very knowledgeable of the organization's business • Stays ahead of current trends such as ISO 9000, TQM, & supply chain management • Knows the theory & practice of marketing, finance, operations, accounting, MIS, & organizational behavior • Is thoroughly familiar with the needs of external customers & their impact on internal customers & suppliers	• Is able to identify problem • Is able to solve problem & not get caught up in the symptoms • Is good & fast decision maker • Is able to think in the abstract • Is able to analyze various strategic options & their direct potential & indirect impact on the organization	• Works well with department members, other departments & suppliers • Is able to reach workable compromise • Can handle conflict • Is adept at organizational politics • Is able to maintain positive mental attitude • Coordinates purchasing departments role with that of the organization	• Knows how to plan materials, budgets, work, etc. • Knows how to organize the department's work & people • Is a good leader • Communicates well & is able to articulate the department's needs to top management and the rest of the organization • Is able to establish a work environment that enhances department members' motivation • Controls well—able to monitor & adjust department activities to meet plans • Can develop purchasing objectives & strategies that are congruent with organizational goals & strategies • Is able to participate in new product/service development, organizational growth, & strategy formulation

cited are the managerial skills and duties of planning, organizing, directing, motivating, controlling, and evaluating purchasing.

Continuous improvement, supplier base reduction, EDI, new supplier relationships, cost and time reduction efforts, and the need to focus on the value-adding strategic dimensions of purchasing have major staffing implications. Progressive organizations often rely on university graduates in disciplines such as engineering, science, business, and law to help them transform the supply function into a comprehensive advantage. It is important that the supply function be staffed with highly competent individuals who can gain the respect and cooperation of all parties in the supply chain, from customers to suppliers. Unlike the past, when most purchasing professionals spent their entire careers within the profession, these high profile individuals may choose to move into other management tasks inside or outside the organization. Such movement of purchasing personnel is a good sign for the profession, because it contributes to the breadth of key individuals who will shape the profession, and it will necessarily contribute to purchasing's reputation in wider organizational circles. Similarly, purchasing should expect, if not solicit, career broadening moves by personnel from other functional specialties early in their careers to effect an early cross-flow of information between functions.

The Impact of Government on Employment

Federal, state, or provincial regulations may also impact the process. During the 1960s and 1970s, legal as well as social changes produced major impacts on an organization's policies and practices in administering personnel functions. In effect, the federal government created an umbrella of regulations that surround and permeate all the personnel programs and practices discussed to this point. The major thrust of these legal interventions has been the prevention of discrimination against various groups of employees. Key regulatory actions are the Federal Civil Rights Act of 1964 (specifically Title VII), Executive Order 11246 in 1965, the Age Discrimination Act of 1967, and the Equal Employment Opportunity Act of 1972.

Title VII of the 1964 act makes it illegal to discriminate among employees on the basis of race, religion, color, sex, or national origin. It focuses on discrimination with regard to any employment condition, including hiring, firing, promotion, transfer, compensation, and admission to training programs. The original act gave the Equal Employment Opportunity Commission (EEOC) the responsibility to try to effect compliance by means of conferences with the groups involved, conciliation, and persuasion. If these efforts failed, EEOC could ask the Attorney General to bring suit against the organization.

In 1972, the Equal Employment Opportunity Act (EEOA) was passed to provide a series of amendments to Title VII, expanding its coverage and strengthening its enforcement. Coverage was extended to virtually all firms with 15 or more employees. The EEOC was given the power to file lawsuits directly against offending firms.

Affirmative Action

Another significant provision of the EEOA requires organizations to take "affirmative action" to move toward achieving a work force that accurately reflects the composition of the community. In other words, an organization must compare its employment, oftentimes by department and by job level, with data on the availability of talent in the relevant labor market. Then, in each case, it must attempt to achieve a work force composed of approximately the same percentage of the various minority groups (including women) as exists in the available labor market. The enactment of this provision is an extension of Executive Order 11246, issued in 1965, which required firms doing business with the government to prepare a written affirmative action plan to accomplish this same result.

The development of a sound, written affirmative action program is seen as a way for an organization to fulfill a social responsibility. In addition, some firms today view affirmative action as a preventive approach to minimize the possibility of problems with the EEOC and also possible subsequent legal actions. Such a plan can be developed using a three-step approach:

1. Determine the jobs in which any group is underrepresented and assess the availability of that group in the labor market.
2. Set numerical integration goals for increasing the representation of that group in those jobs.
3. Specify the actions to be taken to attain the goals.

On July 26, 1990 President George Bush signed the Americans with Disabilities Act (ADA). On that occasion he stated:

> This Act is powerful in its simplicity. It will insure that people with disabilities are given the basic guarantees for which they have worked so long and so hard. Independence, freedom of choice, control of their own lives, the opportunity to blend fully and equally in the rich mosaic of the American mainstream.

The ADA covers nondiscrimination in five major sections, as follows:

Title I	Employment.
Title II	Public service and transportation.
Title III	Public accommodations and service.
Title IV	Telecommunication services.
Title V	Miscellaneous provisions.

We do not have space to discuss all aspects of the ADA, but managers should familiarize themselves with provisions that relate to their type of organization.

Certainly most purchasing managers will have the need to employ people. Below are the major points from ADA Title I—Employment. All organizations with 25 or more employees had to comply with these provisions starting July 26, 1992. On July 26, 1994, the Title I employment provisions took effect for employees with 15 to 24 employees. The requirements state that:

- The ADA prohibits discrimination by an employer against a qualified person with a disability in: recruiting, hiring, upgrading, promotion, award of tenure, demotion, transfer, layoff, and termination.
- All job descriptions should be reviewed to identify "essential functions" to represent an accurate reflection of the responsibility of the position.
- The reasonable accommodation provision appears in Title I in reference to employment of persons with disabilities.
- Requests for accommodations are generally the responsibility of an applicant or employee.
- Employers are required to inform applicants and employees of their rights under the ADA through posting of public notices, job application forms, personnel manuals, etc.
- Undue hardship and reasonable accommodation policies, procedures, and practices should be reviewed to alleviate barriers to employment.

In most organizations, responsibility for developing policies and procedures for complying with EEO regulations is assigned to the personnel department. An effective purchasing manager, however, should initiate actions to work closely with the personnel department in this regard, simply because his or her people management activities must all be designed to fit within the framework of these legal requirements.

PROMOTION

The issue of hiring someone to fill a job vacancy is closely connected to promotion, since someone capable may already be working in the organization. Professional certification, career advancement, and promotion from within are discussed next.

Standards/Certification

Certified Purchasing Manager (CPM) is the designation accorded by the National Association of Purchasing Management (NAPM) to those who apply and pass rigid professional requirements. The Canadian equivalent is the Certified Professional Purchaser (CPP) designation, which is administered by the Purchasing Management Association of Canada (PMAC). Such criterion of professionalism denotes a standard of respectability. For the individual, certification assures peer recognition, better job opportunities, enhanced value to the employer, and quicker professional advancement. For the employer, it states that the purchasing professional has met certain standards deemed important by the profession and has acquired the skills to do a good job. It often allows the employer to remove loose, haphazard, or arbitrary promotion practices and can allow the establishment of sound criteria of knowledge requirements for advancement. It enables the employer to realize the greatest potential from the purchaser.

Career Advancement

Most purchasing departments are relatively small, usually employing from 0.5 to 1.5 percent of the total number of employees in the firm. In addition, purchasing employee turnover typically is low. Hence, advancement of personnel through the ranks, although steady, does not occur rapidly in most established companies. Nearly every purchasing department strives to hire and develop a number of ambitious, promotable personnel to build a creative, management-oriented organization. However, not all expediter's, assistant buyer's, and buyer's jobs should be filled with individuals who will become readily promotable. Usually, there are not enough vacancies into which all can be promoted. And when a person's job no longer offers a challenge, he or she soon becomes discontented and either leaves the organization or performs in only a mediocre fashion. For these reasons, it is essential during the hiring process to match an individual's qualifications both to current and to anticipated future job requirements of the department. It is much easier and less costly to weed out marginal performers before, rather than after, hiring has occurred. An effective purchasing individual properly trained is invaluable to the firm; a poor one is a major liability.

Advantages/Disadvantages of Promoting from Within

Purchasing personnel at all levels can be acquired either from within the firm or from external sources. When vacancies occur above the expediter's level, the common practice in most companies is to promote personnel from within the department. Promotion from within produces several distinct benefits. First, the practice tends to keep morale high because employees know that they are not "trapped" in dead-end jobs. Such actions stimulate individual performance and growth by offering an avenue of advancement for the better career employees. Second, promotion from within reduces total training costs. Often a minimum of training is required because the individual's past experience in the organization is generally useful in the new position. If a person is hired from the outside to replace the promoted individual, he or she can thus be brought in at a lower level, which inevitably also reduces training requirements.

However, promotion from within can also lead to problems. One promotion may result in a chain of lower-level promotions, simultaneously moving several people one step up the organizational ladder. If chain promotions occur frequently, the organization tends to lose its stability, because a large number of individuals are continually learning new jobs. When a company is growing rapidly, this policy sometimes results in the promotion of people who are not ready to be promoted (the Peter Principle). The mediocre performance resulting from such action simply compounds the problem of instability. Finally, promotion from within produces "inbreeding." If carried to extremes, it may jeopardize the flow of new ideas into the organization.

Advantages/Disadvantages of Hiring Outside Employees

A wise manager promotes from within when it is practicable. When such action tends to generate problems, however, personnel should be drawn from external sources. This latter practice has considerable merit, particularly in the case of special staff jobs, like purchasing, because it brings new ideas into the organization. It also prevents substituting seniority for ability.

When a purchasing manager must go outside the department to fill position vacancies, such personnel may be drawn from other departments within the firm. Transfers of this type usually occur at the lower levels and may produce significant advantages for the purchasing department. In the first place, the person transferred is familiar with company operations and can usually assume full job responsibilities sooner than a new employee. Second (and more important), such a transfer brings into the purchasing department experience in a related functional area. A person coming from an associated area has experience that may be useful in buying activities and may also provide a strong liaison with user departments.

Succession Planning

Management succession planning is considered to be the sum and substance of human resources planning. It refers to plans that organizations make for replacement of their key executive personnel. In some companies, this can involve the development of a management replacement chart, which lists, for each position in an organizational chart, the name of the current holder of the position, and the names of one or two replacements for each of those positions. Another method is to develop a prediction of each individual's expected job in five years and at the end of his or her career.

ORIENTING NEW EMPLOYEES

Employee orientation involves introducing new personnel to the job and, more important, to the organization. The applicant receives some orientation to the new job during the hiring process, but frequently this exposure serves only as a superficial introduction. The new employee needs a much more formal and complete orientation. An employee will want to know what is expected of him or her on the job. It is the responsibility of management to ensure that all employees know what is expected of them.

An orientation program should be designed to relieve feelings of insecurity in the new environment. The employee should be made aware of the organization's history, products, and operations. If not communicated during the interviewing process, the organization's values and visions must be shared and embraced by the new purchasing professional. During this time frame, questions about culture and goals should be resolved. Such orientations often include formal instruction, personnel manuals, employee handbooks, and tours. Usually, orientation programs are run by human resources departments, but the new employee's supervisor and/or purchasing manager will also play a major role. It is important to remember that the early exposure a new employee receives to the organization will have a resounding impact on his or her early development on the job.

Initial Job Training (Functional Orientation)

If a new employee is expected to achieve a desired level of productivity in a reasonable period of time, a certain amount of job training is inevitably required. On-the-Job Training (OJT) is still the most common way of training new employees in business, industry, government, and medicine. The problem is how to make OJT as effective as possible. The biggest reason that OJT is not as effective as it could be is that it is not planned. To make OJT productive, it is necessary to outline those aspects employees are expected to learn for each job classification. Subsequent- follow through with the individual trainers *and* the employees to ensure that the employees learn what is expected is also important. There are four major approaches to initial job orientation:

1. ***Learn by doing***—Also known as "trial by fire," the most basic method of training new employees is to give general guidelines as to what needs to be done and to have the employees teach themselves the nature of the job. In fact, many new employees learn their jobs through this process. Of course, this method cannot be employed in all instances and should be used with caution. A likely scenario for employing learning-by-doing is when the employee is already well trained or experienced in the work he or she has been hired to perform, and no peer, supervisor, or training facilities are available as knowledge resources. In such a case, the new employee is left to his or her own devices, and has to rely on intuition and experience as to how the job should be performed.

2. ***The sponsor system***—A commonly used practice involves assigning a sponsor (or a "buddy") to the newly hired person. After the supervisor has provided adequate initial orientation, the new employee is assigned a specific job in which he or she also receives a certain amount of basic initial job instruction from the same supervisor. The bulk of the training load, however, rests with the individual's sponsor, who is an employee doing similar work. The sponsor acts as an informal trainer during the entire period the new person is learning the job. A sponsor should be chosen for his or her experience and, more important, teaching ability.

This approach can be quite effective if the sponsor is a good teacher. It has the probable disadvantage, however, of restricting the new person's initial training to a single job. Some time may elapse before he or she realizes fully the many implications of the activities of the job and how they relate to departments outside purchasing. Also, the time devoted to such training activities may significantly reduce the sponsor's productive output.

3. ***Formal classroom training***—In some organizations, initial job training consists of a series of brief classroom courses dealing with theoretical principles underlying purchasing and related activities. Such programs prepare the new purchasing employee to do a better job and also aid him or her in establishing rapport with personnel in related departments.

4. ***Functional rotation***—To minimize the shortcomings of the "buddy system," many companies modify it by adding an element of functional rotation training. Before a new employee is assigned to a specific job, he or she is considered as a trainee for a period varying from several weeks to several months. Much of this initial training is spent in departments other than purchasing. The basic idea is to expose the individual to a number of functional activities both within and outside the purchasing area and to facilitate his or her understanding of the various purchasing functions and their relationships to other operating functions. A typical program includes assignments in such areas as receiving and stores, purchasing records, expediting, assistant buying activities, inventory control, and selected line production departments. Specific assignments vary depending on the person's background, and on his or her first permanent job assignment. The program objective, however, is to develop a general understanding of the complete material cycle in the firm. Upon completion of rotational training, the new employee is assigned to a specific job where he or she may receive further job training from a sponsor or from the supervisor.

Continuous Professional Development

If a manager expects to utilize employees effectively over the long term, he or she must assume the responsibility for assisting and guiding them in the continued development of their capabilities. Determining an individual's specific development needs is a product of observation and periodic counseling by the manager of the individual employee. These needs should be jointly determined, and plans for subsequent training experiences should also be planned jointly for the ensuing six- to twelve-month period. A management-by-objectives (MBO) program provides an unusually effective opportunity for the cooperative determination of such an individual development program. There are many options and combinations of alternatives to assure continuous professional development. These include:

1. *Job rotation*—"Job rotation training" can take several forms. As the name implies, it involves a process in which the purchasing employee is rotated from one job assignment in the department to another, until the employee develops reasonable competence in each of the jobs. The objective is to provide an expanded base of purchasing department competency, as the individual's development permits.

Following experience as an expediter and subsequently a junior buyer, an individual's first fully fledged buying assignment typically is in the MRO section or in the general purpose material section. From this position, the individual may elect to specialize in buying a particular type of raw material or production component. Such job progression training thus provides the purchaser with experience in various types of departmental activity, each with increasing complexity. Although each production buyer typically becomes somewhat of a skilled specialist, his or her capabilities can be expanded further with experience in several specialized buying jobs. This type of rotation naturally occurs slowly because of the extensive background and knowledge required in each buying area.

Conducted wisely, this technique not only fosters professional development of personnel but also provides buying flexibility within the department. Care, however, must be taken to not dilute technical buying competence by rotating buyers too frequently.

2. *On-the-job training*.—Most "on-the-job" training for continuous professional development is conducted informally. It can be initiated by various members of the purchasing department in response to the manager's observed needs among departmental personnel. Such training may consist of simply supervisory coaching for selected individuals. In other cases, it may take the form of periodic discussions among management and selected groups of personnel. Each session is conducted in seminar fashion and focuses on the exchange of ideas about relevant purchasing topics. On-the-job training may also include periodic lectures and demonstrations. Some firms periodically hire consultants to conduct refresher sessions on such topics as value analysis, cost estimating, negotiations, and similar practical purchasing skills and techniques.

3. *Self-training*—As noted previously, responsibility for recognition of specific needs lies with the individuals themselves, and with their supervisors. For this reason, much of a buyer's professional development is acquired through carefully directed "self-training." Self-training takes many forms, but commonly involves the following approaches:

(1) Studying purchasing periodicals, books, and research reports.

(2) Studying selected business publications.

(3) Studying trade magazines and special resource books on materials.

(4) Attending purchasing association meetings and special commodity group meetings.

(5) Enrolling in selected university or community college evening courses and correspondence courses.

(6) Attending special seminars and workshops sponsored by universities, community colleges, and professional societies.

During the course of an individual's self-training endeavors, one of the manager's responsibilities is to help the individual attain a balance between the development of business skills and technical knowledge of materials.

4. *Site visits to organizations/suppliers*—Training of purchasing personnel should include visits to suppliers. From such visits, trainees obtain an understanding of the suppliers' capabilities as well as an enhanced industry knowledge. With such information, the trainee can evaluate the potential use of each supplier's firm for his or her company's business.

5. *Formal/classroom training programs*—Classroom-type training, or *formal training,* may not actually take place in a classroom. It could be done by calling the employees together at the back of the store or in your office or by a work station. What distinguishes formal training from the other types is that it is carried out by means of typical school teaching techniques, such as lectures, audiovisual aids, case problems, role playing, or discussion. Formal training is usually done by the purchasing manager or the training director, or even by a specialist who has been brought in from outside the division or company.

Formal training has the advantage of allowing more information to be covered in a shorter time. It affords greater control over what is being taught so that everyone learns the same thing. It also allows employees to learn from colleagues. A disadvantage of teacher-centered training is that it is rarely as effective as learning by doing. Therefore, employees tend to retain less than they do when the learning takes place through OJT. However, if the concepts are continually reviewed, key concepts are often retained longer.

Case-based learning using empirically developed teaching materials also offers a significant opportunity for many organizations. Experiential learning has been shown to be very effective in managerial development, particularly in the purchasing arena.

Many organizations have in-house training programs that are provided by professional training organizations, NAPM or their affiliated local associations, colleges and universities, and the American Management Association. They often employ combinations of films, lectures, case studies, and programmed instruction.

Smaller organizations that lack the funds needed for formal training programs often turn to outside agencies and programs, such as the seminars run by NAPM and the AMA. A major advantage of outside instruction is that it exposes the participant to colleagues from a wide variety of organizations. However, these programs, by necessity, tend to be general in order to meet the needs of a broad scope of purchasing participants.

6. ***Peer-to-peer***—Purchasing professionals can learn an enormous amount from their peers in the office who have more training and experience in different areas. An atmosphere of cooperation and harmony (and not competition) is necessary for such interaction to take place.

7. ***PHRASE***—NAPM's PHRASE system is a three-step program designed to enable firms to target specific training and development needs for their purchasing personnel. The program begins with a job analysis that identifies the important tasks of the purchasing department in question. Next, a diagnostic test is given each participant, which measures his or her comprehension of the basic tasks surveyed in the job analysis. The examinees are compared to nationwide norms to determine their strengths and weaknesses in the various purchasing tasks. A third phase combines the results of the job analysis and the diagnostic to create a customized training program, targeted to the areas that are important to the organization but not well mastered by the staff. In this way, the training programs are designed to maximize the performance of the personnel in the most efficient way possible.

8. ***Media***—Many textbooks and publications (e.g., *NAPM Insights, International Journal of Purchasing and Materials Management, Purchasing,* etc.), plus audio and videotapes distributed by NAPM and other associations, are available to purchasing personnel for their professional development and career enhancement.

What is Retraining?

Retraining, of course, means training again. There are two reasons for retraining someone. First, a person who is trained to do one job may get transferred to a new job. Second, an individual may be doing the job or parts of the job incorrectly. Retraining of this type is often more difficult than training someone new; it is hard to break a bad habit and to teach a new way of doing things. It may be especially difficult to motivate someone who has been doing things incorrectly for some time. Start retraining an employee who is doing a job wrong by explaining why the job should be done differently. Try to get the person to understand why it is important to learn the new method you are going to teach. From this point on, the retraining process is just like training, except that it usually requires greater effort, with more repetition and follow-up.

DEVELOPING TRAINING MANUALS

Developing a comprehensive training manual is the responsibility of every purchasing department. Staff training manuals and policies and procedures manuals can be used to train personnel. New or untrained personnel may be overwhelmed with purchasing job requirements. Manuals can be referred to as questions of policy or procedure arise during and after orientation and initial training. Procedures manuals are especially useful where extensive details in routine operations are needed. These manuals make supervision easier, encourage standard practices, improve procedures, and aid in training.

TRAINING, COST EFFICIENCY AND MEASUREMENT OF OUTCOME

The outcomes of training should be measured in behavioral and operational terms to determine the effectiveness of the training effort (including how trainees actually behave on their jobs) and the relevance of their behavior to the organization's objectives. In this way, one can assess the utility or value of the training. The types of questions that need to be answered when evaluating training programs include the following:

1. Was there a change in knowledge, skills, and/or abilities related to purchasing effectiveness in the various participants?
2. Were these changes due to the training?
3. Are the new skills positively related to the organization's goals?
4. Will similar changes occur for new participants in the training program?
5. Was the training cost effective? Was it worth the expense to the organization? Could the same effect have been achieved with another less expensive training mode?

To determine the answers to these questions, techniques of educational research, including tests, questionnaires, interviews, and experimental design, need to be employed by trained researchers. Using objective quantitative and qualitative research methods, the evaluators can answer the above questions and provide data for making decisions on whether to continue, discontinue, or modify a training program. Only in this way can one make an objective, informed assessment of a training mode.

The Purchasing Manager's Role in Training

Since most training is OJT, the purchasing manager is responsible for the majority of training that takes place in any organization. This allows the manager to get to know the new employee, and vice versa. If there is not enough time for the manager to conduct the training individually, it is preferable to rely on an experienced member of the department to train the new employee. However, it is important to choose an employee who is a good trainer, who likes and respects the manager, and who has a high opinion of the company.

TRAINING, RETRAINING AND COACHING

Training Tracks

When contemplating the training of purchasing professionals, it is helpful to divide the content into three training tracks or divisions: (1) product knowledge; (2) the organization's own policies, procedures, and methods; and (3) purchasing principles.

Product Knowledge

People who have been in purchasing for a while become highly knowledgeable about the products they buy. Experienced purchasing people are most often hired because of their product knowledge. In fact, typical purchasing job announcements usually include a headline that reads: Steel Buyer Wanted, Electrical Buyer Wanted, or Capital Equipment Buyer Wanted.

If, on the other hand, buyers need additional product knowledge, it is certainly possible to structure such a program. However, of the three training tracks, product knowledge is the most difficult and time consuming to establish. Different buyers need different product knowledge, since not everyone in the department buys the same items. To develop such programs or to conduct the training, many organizations use consultants or educators who are experts in purchasing principles training.

Organizational Practices

Even though you may have experienced department people, someone should review the most important policies, procedures and methods with the entire department at least once a year. People who know what they should be doing, do not always follow the rules. Reviews of proper practices have a tendency to get everyone back on the right track. Airline pilots have to repeat the standard operating procedures every six months to make sure that they have not developed any bad habits.

The best method to use in making such a review and teaching those who are new seems to be lecture and/or discussion. Usually, the right person(s) to conduct this type of training would be one, or a combination of, the following: (1) purchasing manager; (2) a representative from the accounting, engineering, quality control or legal departments; or (3) someone from training.

Purchasing Principles

As a rule, of the three tracks, the area of purchasing principles is the one in which purchasing people need the greatest improvement. Strange as it seems, this is even true of people who have been in purchasing for a long time—if they have not had formal training in the field.

As far as content goes, your best source, as a starting point, is the C.P.M. study guide and the text for the C.P.M. review. Review these and choose subjects that seem to fit the needs of your organization.

The best methods of instructing people in purchasing principles include: (1) lecture; (2) case problem; (3) role playing; (4) audio-visuals (TV and movies); and (5) research projects that end with the practical implementation of the findings.

Who will be doing the instructing? Here, we will run the risk of being accused of making a self-serving statement, but we honestly believe that this is where you should consider bringing in a consultant or educator who is an expert in purchasing principles training. The reason we say this is that very few of even the Fortune 100 companies have competent people on their staffs who can do this kind of training. In fact, there are only a few in the entire country—which is unlike management, sales or accounting training, for the woods are full of very competent people who can do this type of training.

RESOLVE EMPLOYEE PERFORMANCE PROBLEMS

Most organizations have work rules that, if violated, can result in various penalties, including dismissal.

Corrective Action Process

The principle of corrective action (sometimes called progressive discipline) means that management responds to a first offense with some minimal action, but to subsequent offenses with more serious penalties such as layoff or discharge.

1. ***Notice of the problem***—The first step in the corrective action process is to warn the employee of the problem. Usually, this is first done orally, and if the problem continues, it is then given in written form stating the consequences of future offenses.
2. ***Effort to resolve the problem***—The employee must be given sufficient time to demonstrate a change in behavior. This may be over several months, depending on the problem.
3. ***Monitoring progress***—The manager should carefully monitor the progress of the employee, and offer feedback, especially if the behavior change is positive, to give the employee a sense of whether or not his or her actions are appropriate.

Types of Problems

1. ***Performance problems***—Performance is probably one of the main employee work problems. This may include frequent errors, oversights, sloppy work, or slow performance (i.e., failure to meet deadlines). It can also include habitual lateness or missed days.
2. ***Attitude, stress, burnout***—A great many employee performance problems can be linked to poor attitude, due to on-the-job stress and burnout. Many organizations have in-house counselors to deal with troubled employees, to help them change their attitudes toward their positions and to develop a positive approach.
3. ***Theft***—Incidents of employee pilfering, embezzlement, or theft are serious violations that may involve criminal proceedings, as well as disciplinary action or dismissal.
4. ***Substance abuse***—Estimates of the percentage of workers affected by alcoholism run from 5 to 10 percent. A somewhat smaller percentage have problems with drug abuse. Many employers offer counseling and treatment for substance abusers, an expense frequently covered by employee benefit packages.

Documentation

In all cases, incidents of employee difficulties should be well documented by the manager, including dates of incidents, warnings, meetings with the employee, and the nature of all manager-employee discussions.

Collective Bargaining Requirements

Collective bargaining is the process of negotiating a labor agreement between union representatives and employer representatives. The contract negotiated between employer and union representatives is called the collective bargaining agreement. It sets forth the terms and conditions under which union members offer their services to an employer. Such agreements usually have specific requirements for dealing with employees who are performing poorly. It is the responsibility of the manager to ensure that those requirements are met whenever he or she engages in disciplining or dismissing an employee.

Employee Assistance Programs

Employee Assistance Programs (EAPs) are formal counseling programs aimed at assisting employees with emotional problems due to such difficulties as personal or family crises, substance abuse, or emotional illness. Typically, such programs reside in or are connected to the employee medical department. Such programs can assist in the general mental health of an employee and can also help improve communications and understanding between superior and subordinate.

EMPLOYEE APPRAISALS

Determine Appraisal Factors

How does a purchasing manager determine exactly which factors of an employee's performance should be evaluated? Commonly used factors are:

1. Quality of work.
2. Quantity of work.
3. Knowledge of specific job activities.
4. Cooperativeness.

5. Dependability.
6. Initiative.
7. Accuracy.
8. Adaptability.
9. Attitude.

Each purchasing department must choose the factors that are related to its specific job responsibilities.

Conduct Interviews/Give Feedback

Performance reviews are usually conducted as face-to-face interviews of the employee by the purchasing manager. Managers often dislike this sort of activity. The reasons for this include the following:

1. The manager's appraisal may often differ with the subordinate's self-appraisal.
2. It is difficult for most managers to carry out the appraisal in such a way that the reaction of the employee is positive, since supervisors often lack the necessary interpersonal skills.
3. Raters often have a distaste for critical analysis of others' performance and for the conflict that sometimes results from a performance review.
4. Managers get little positive reward from the entire performance review experience.

Purchasing managers should keep in mind that there is a tolerance limit for criticism. Most individuals probably feel they have some weaknesses or deficiencies and have not performed as well as they might have in all instances. However, when the criticism exceeds what is perceived to be valid, the "critical level" is exceeded, and the individual may exhibit anger or tension. In such cases, communication difficulties compound rapidly, and the chances for obtaining a useful outcome diminish.

For feedback to be effective it should be as precise as possible. Managers should not speak in general terms, because what they are saying could be interpreted in a number of ways. Negative feedback should be timely and be given shortly after the incorrect performance or behavior has taken place. Also, feedback should be impersonal. Criticizing personal traits is especially likely to cause an emotional reaction, and as a result, the nature of the performance deviation itself may be overlooked. Feedback should be obvious and clear and be provided early and often, so that the individual is made aware of the information and can take corrective action when required. Understanding performance is enhanced when there is frequent review and when feedback is received early enough to identify problems in goal achievement.

Team and/or Peer Input

At times, feedback and review come in the form of team or peer appraisals. These approaches have been used for many years in the military and in academia, as well as in industry. The unique first-hand knowledge that peers often have of each other's performance is the rationale for the use of such appraisals. However, if conditions of competition exist, this type of rating may not be very effective.

Self-Assessment

The advantages of self-assessment by employees are that employees are in a privileged position to evaluate themselves with respect to job knowledge and performance frequency. Such assessments also foster development and lead to less defensive appraisals. However, some employees have a tendency to inflate ratings when making self-assessments and may attribute their problems or poor performance to their environment or to others, not to themselves.

Employee Accountability

A good performance appraisal identifies those areas in which an individual has problems and specifies what the employee needs to do to correct those problems. The appraisal thus provides the employee with guidance on how to improve, and makes him or her accountable for that behavior, with a definite time span for improvement.

Customer Input

At times, the purchasing manager may be able to obtain useful performance appraisal data from the user departments with which the buyer has frequent contact, to see if their needs are being met promptly, courteously, accurately, and with minimal problems. Purchasing may also desire such information from members of the supply chain, although it may be very difficult to incorporate such input without leading to potential conflicts of interest.

USES OF PERFORMANCE APPRAISAL DATA

Salary Increase

No department operates at its full potential for long if its salary structure fails to reward individuals in relationship to their respective performance levels. A good performance appraisal program does not guarantee an equitable salary structure. It does, however, provide data that can be used in developing a sound compensation plan or in correcting an inadequate one.

Promotion

How do managers know which people in their departments are likely to become candidates for the top jobs? They determine this by analyzing each aspect of an individual's performance record. It is imperative that such analyses be made using detailed and accurate written data. A well-designed appraisal program provides the required data.

Personal/Career Development

The most important benefit that can come from a good employee evaluation program is the information needed to stimulate and direct each individual employee's professional development. A supervisor's prime responsibility is to develop capable and effective personnel. The data provided by appraisals can be analyzed to determine each employee's strengths and weaknesses. This determination facilitates the development of a realistic professional improvement program for each individual.

Employee Morale

Every purchasing manager must develop a carefully structured program for appraising the performance of personnel. Nothing is more disastrous to the morale of a department than haphazard and inconsistent evaluation of employees' performance.

Disciplinary Action

A well-designed employee performance appraisal is a necessary guide to disciplinary actions that are focused, fair, and provide direction for employee improvement.

Employee Recognition

Purchasers (as well as other staff) want to know that their diligence will be repaid. When departments reward their hardest workers, heavy workloads are viewed as an opportunity to win recognition, earn respect, and be included in the most interesting and high profile projects. Most purchasers are motivated when their managers make them feel good about quality work by acknowledging their superb efforts. Many organizations find that programs that allow purchasers to be recognized and regarded for their performance benefit the entire organization.[3]

ISSUES IN THE TERMINATION OF EMPLOYEES

Hiring the wrong person occurs periodically despite attempts to the contrary. Although this text describes ways to avoid such occurrences, it happens to all managers occasionally. In such a situation, the best response is to recognize and admit the mistake early and to deal with it swiftly. Before dismissing an employee, one should investigate the possibility of upgrading the employee by training or development. Developing a successful buyer is a long-term investment, and behavioral modification that focuses on changing one's personality in order to make an ill-suited match is often doomed to failure. A replacement could be just as bad or worse, and it is costly to hire new employees. Also, an organization that fires a large number of employees gets a bad image; often, potentially successful workers may be reluctant to apply for a job there. Still, if training obviously will not alleviate the situation, then the only choice may be to terminate the employee.

Consistent Documentation/Evaluation

Employees should never be terminated on a "whim" or for vague, personal reasons. Generally, employees should be terminated for consistently poor performance, insubordination, serious violations such as theft, or chronic substance abuse. The employee should be given a reasonable amount of time to improve the poor performance or behavior (except in the case of theft). In all

---

[3]For further reading about employee recognition, please see Harris, G., "Pay for Team Performance: Gainsharing," NAPM Insights, November 1992, p. 8, and Sunkel, J., "Incentives to Purchasers," NAPM Insights, September 1991, p. 26.

cases, the reasons for termination needs to be documented and the employee made aware, before actually being fired, that he or she is on probation.

Personnel Policies, Procedures, and Union Requirements

Most companies have well-outlined policies for the termination of employees. In the case of union employees and layoffs, union contracts usually specify a laid-off worker's right to be recalled based on seniority.

KEY POINTS

1. A purchasing department's staff directly impacts on how effective the department will be. Care must be taken when developing job descriptions and responsibilities.
2. Regardless of the nature or level of a buyer's position, desired employee characteristics include product/technical knowledge, and decision making and negotiation skills, with interpersonal skills expected to become the most important in the future.
3. A purchasing manager must possess all the skills of both a buyer and a manager, as well as the ability to analyze problems and think strategically.
4. The continually changing role of the purchasing department has brought about a new, educated, mobile purchasing professional.
5. Care must be taken by a manager to adhere to and promote the social and regulatory requirements of the community and the state when hiring for a purchasing department.
6. Job vacancies can be filled from internal, external, or predetermined sources.
7. When selecting for a position, standards and certifications can help to prequalify a candidate.
8. Promoting from within ensures that ambitious employees stay motivated and continue developing.
9. Hiring from outside the organization provides the purchasing department with fresh new ideas and concepts, but requires higher training costs and longer employee development times.
10. An effective orientation plan can reduce a newly hired employee's anxiety. Peer support, corporate literature, and

managerial interviews help accelerate the employee's early development.

11. Subsequent training helps new employees to become productive. Learning by doing allows a new employee who already has experience in the position to leverage their skills and develop their own methods. The sponsor system shares the experience gained by a co-worker with the new employee. Formal classroom training provides a good theoretical background for the employee. Functional rotation exposes the employee to the various departments of the company, and helps to develop a better overall understanding of the company's operations.
12. Employee development is an important responsibility for both the manager and the employee in ensuring long-term productivity.
13. Job rotation gives the employee an expanded knowledge of the whole purchasing department and its operations. The department gains flexibility and the employee gains experience.
14. On-the-job training focuses on key ideas and topics, such as value analysis and negotiations.
15. Self-training, through books, magazines, seminars, and courses, provides much of a buyer's development.
16. Formal training programs tend to be general and expose employees to a wide variety of organizations and issues. NAPM's PHRASE program enables a firm to focus training on the specific needs of its purchasing employees.
17. Although the retraining process is much the same as regular job training, more effort is needed in explaining the importance of new methods or procedures.
18. Most employee problems can be categorized into four areas: performance-related problems, attitude/stress-related problems, theft, and substance abuse.
19. Progressive discipline deals with these problems by first placing the employee on notice. The employee is then allowed sufficient time to rectify the problem. Progress is monitored and feedback is given to the employee. This process, and any employee problems, should always be well documented.

20. Performance review interviews can be used effectively to provide feedback and constructive criticism to employees. Comments should be focused, impersonal, and without bias.
21. Team/peer input, self-assessment, employee accountability, and customer input are all sources for useful performance appraisal data. This data can indicate to an employee areas or problems that need to be corrected.
22. Performance appraisal data should be used in the development of a compensation plan, in determining candidates for promotion, in employee career development, and during disciplinary situations.
23. When retraining or development programs are ineffective in dealing with under-performing employees, the probation and termination process should be well documented and communicated to the employee.

CHAPTER 8

DIRECTING: LEADERSHIP AND COMMUNICATION

PRINCIPLES OF DIRECTING

Directing

Directing is the use of communication and leadership to guide the performance of one's subordinates toward the achievement of the organization's plans. Some authors describe the process of directing as being one of leadership, communication, and motivation. This text, however, does not include motivation as part of directing, in order to give motivation more emphasis. So the next chapter deals exclusively with motivation as a function of management, while this one treats directing as a process of leadership and communication. This chapter discusses how the purchasing manager can take the plans they have made and the organizing they have done for the purchasing department and put them into practice.

Two Basic Principles of Directing

Both leadership and communication depend on two basic principles of directing. The first principle is that of directing toward objectives: *A primary responsibility of management is to direct the employees toward well-defined and well-known organizational objectives.* To direct employees, one must first answer the question, "Direct them where?" A plan with well-defined, measurable objectives and target dates is required. An example is: to increase average on-time deliveries by 10 percent this year.

The second principle of directing is the principle of unity of purpose: *An organization functions best when the organization's objectives and the individual's can be made to coincide*. The result is an efficient completion of the organization's goals and a fulfilling experience for each individual. If employees are told to work hard so the company will make a profit, they probably will not. But if they are shown how working hard will result not only in profits for the company but also in a bonus, promotion, secure job, or other benefits for themselves, they are much more likely to work hard.

LEADERSHIP

Definition of Leadership

Good leadership is the ability to inspire others to move enthusiastically toward the organization's goals. Good leaders know where they are going and can get others excited about going there with them. There is some disagreement as to whether leadership is a quality that can be developed or is innate, as the following excerpts show. Peter Drucker, a noted authority on management, feels that leaders must be born rather than made.

> Leadership is of utmost importance. Indeed, there is no substitute for it. But leadership cannot be created or promoted. It cannot be taught or learned. There is no substitute for leadership. But management cannot create leaders. It can only create the conditions under which potential leadership qualities become effective: or it can stifle potential leadership.[1]

Warren Bennis, author of many books on leadership, holds an opposing view.

> Although I have said that everyone has the capacity for leadership, I do not believe that everyone will become a leader, especially in the confusing and often antagonistic context in which we now live. Too many people are mere products of their context, lacking the will to change, to develop their potential. I also believe, however, that anyone, of any age and in any circumstances, can transform himself [or herself] if he [or she] wants to. Becoming the kind of person who is a leader is the ultimate act of free will, and if you have the will, this is the way.[2]

---

[1]Drucker, Peter F., *The Practice of Management,* New York: Harper & Row, 1954, pp. 158-159.

[2]Bennis, Warren, *On Becoming a Leader,* Reading, Massachusetts: Addison-Wesley Publishing Company, Inc., ©1989, p.8; reprinted with permission of Addison-Wesley Publishing Company, Inc.

Recognizing the dichotomy of these approaches, the reality is probably somewhere in between. Certain characteristics of leadership are ingrained very early in an individual's development. On the other hand, many attributes of good leadership can be studied, emulated, and adapted by successful managers.

VARIOUS STYLES OF LEADERSHIP

Definition of Leadership Style

Leadership style is the position that a leader usually takes with regard to how much decision-making freedom he or she allows subordinates to have. The amount of freedom can vary from none (oppressive autocracy) to a great deal (participating leadership).

In 1958, Robert Tannenbaum and Warren H. Schmidt published an article in the *Harvard Business Review* called "How to Choose a Leadership Pattern." This has been one of the most popular articles ever published by the *Review*—in fact, over 300,000 reprints have been mailed out. Therefore, the *Review* decided to publish an updated version of the article in 1973.

Tannebaum and Schmidt are not the only ones who have worked on this problem of defining leadership styles. For example, as early as 1944 Kurt Lewin constructed a leadership model that had three styles: (1) autocracy, (2) democracy, and (3) laissez-faire. More recently, Rensis Likert has proposed these four systems of organizational climate: (1) exploitive, (2) benevolent authoritative, (3) consultative (4) participate group.[3]

Using Tannenbaum and Schmidt's work as a starting point, and building to a degree on Likert, what follows is a somewhat simpler analysis of leadership styles. This approach is also an attempt to describe leadership as it is, rather than how it might be; that is, to portray leadership styles as they are generally practiced in purchasing organizations today.

Four Leadership Styles

There are basically four leadership styles: (1) oppressive autocratic; (2) benevolent autocratic; (3) consultative; and (4) participatory. They are depicted in Figure 8-1.

---

[3]For a full discussion of organizational climate, see Rensis Likert, *The Human Organization,* New York: McGraw-Hill, 1967, pp. 14-24, 120-21.

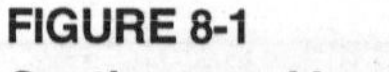

FIGURE 8-1
Continuum of Leadership Styles

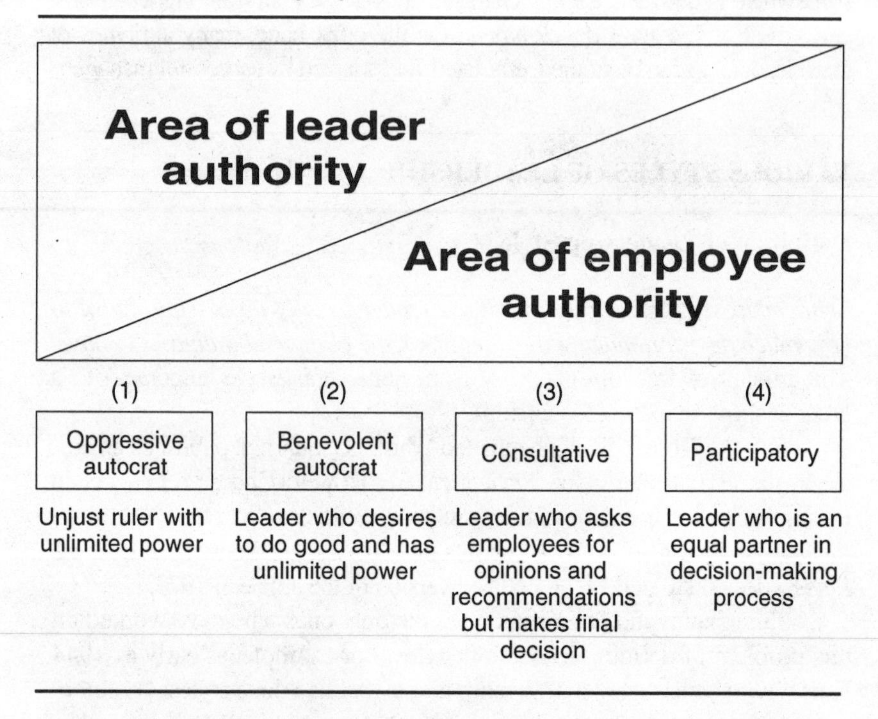

The ***oppressive autocrat*** is an absolute dictator. This leader feels that he or she is in the best position to judge what is good for all and thinks that the only true superior-subordinate relationship consists of subordinates doing what they are told without question. This kind of leadership does not necessarily imply the use of force, but force may sometimes be used. Some employees respond well to this type of leadership, but most do not.

The oppressive, autocratic kind of leadership appeals to two types of managers: (1) those who are untrained and know of no better way; and (2) those who are self-centered and need such a style in order to feel important. We advise you to avoid using this style of leadership—it is a method that has outlived its usefulness.

The ***benevolent autocrat*** is a leader who is both task-oriented and employee-oriented at the same time. That is, while such a leader is primarily concerned with the accomplishment of tasks, he or she does take into consideration the wishes, feelings, and needs of employees. And as far as possible, the benevolent autocrat tries to do what is best for the employees, as well as to keep them informed about the reasons for his or her decisions. Benevolent autocrats may ask employees for facts and information before making decisions but do not ask for their recommendations, nor do they permit them to make decisions or take part in the decision-making process.

The attitude of benevolent autocrats toward employees is typically warm, friendly, courteous, and respectful. They do not regard employees as second-class citizens but rather respect them as equals in the human family. They do not make a show of rank to make themselves feel important and the employee feel humble. But there is never any question about who is in charge.

Consultative leaders suggest a course of action to their employees. Their attitude is that they want the honest reactions of their subordinates and are open to new alternatives. That is, their employees may be able to persuade them to accept their solution. However, this style does not necessarily imply a one-person, one-vote situation. In the final analysis, the leader makes the decision alone. He or she must accept full responsibility for these decisions even if they have been based on bad advice.

In ***participatory leadership,*** leaders involve their subordinates in the decision-making process. They invite alternative solutions to problems and include them with their own ideas. From the collective set, the group selects the solution that seems to be the best. Such leaders do not exercise their authority—rather their vote is equal to those of other members of the group. They accept the decision of the group even if it is contrary to their wishes. Consensus and buy-in are strong points of this style, but the biggest drawback is the time required. It is much faster for one person to make a decision than for many to participate.

DEVELOPING A LEADERSHIP STYLE

The appropriate leadership style depends on a number of conditions such as (1) ability and personality of the leader, (2) ability and personality of the employee, (3) the work situation, (4) the leader's own supervisors (people tend to lead as they are led), and (5) organization practices and policies. In addition to these five issues, four other factors should be considered.

The first factor is that the success of a leader should be judged by *long-term results;* That is, not by how much the organization accomplishes in one day but by how much it accomplishes day after day for a year or more. The second point to consider in developing a leadership style is how to provide *consistency and flexibility* simultaneously. Nothing lowers group morale faster than a manager whom employees cannot anticipate—one who is a dictator one day and seeks participation the next. Good leaders choose a style with which they feel comfortable most often and deviate from it only when absolutely necessary. A temporary move toward employee-centered leadership is generally welcomed, while a move toward task-centered leadership is welcome only if the situation warrants such a change.

Third, it is necessary to recognize that what is important to the employees is not the number of decisions made but the *effect these decisions have on individuals.* A buyer is less concerned about the purchasing manager's decision to have the weekly departmental meeting on Tuesdays than the decision to change the buyer's vacation from July to January. A good leader tries to involve employees in decisions that have direct bearing on them.

The fourth consideration is *orientation.* Is the manager task-oriented or employee-oriented? A task-oriented manager places greater emphasis on completing the work than on the needs of employees and develops ways to drive employees (e.g., motion study, time study, close supervision, etc.). The employee-oriented manager tries to lead employees by thinking of their needs and showing them how their needs can be satisfied along with those of the organization. This approach does not mean that an employee-oriented manager is not also task-oriented; managers must always be concerned with getting the job done. What employee orientation does mean is that the boss is concerned with getting the task completed in a more agreeable way. The leader who tends toward an employee-oriented style will usually achieve better long-term results than the leader who is task-oriented.

EFFECT OF LEADERSHIP ON MORALE

Job morale reflects the employees' attitudes toward their employment. In other words, it is the sum of the feelings, both positive and negative, of individuals or a group toward their work. Morale changes from day to day, even from hour to hour. Consistently high morale is not achieved through salesmanship or manipulation or by having a "morale drive" every three months. It can be achieved and maintained only by good, positive leadership over the long term.

High morale has been linked to high productivity so often that many people believe an increase in morale will bring an automatic and equal increase in production. Researchers have found, however, that high morale does not guarantee high productivity. It is possible, for example, to have a purchasing office where the employees are paid to work eight hours a day but work only four hours. Their morale may be very high, but their production is low. What does seem to be true is that high production for an extended period requires high morale.

Furthermore, the Hawthorne experiments showed that poor working conditions do not necessarily lead to low morale and low output. Infantry soldiers could be eating cold rations and slogging around in a jungle (poor working conditions), yet their morale and productivity might be very high if they believe in the cause for which they are fighting and feel they are making a worthwhile contribution to it.

COMMUNICATION AND LEADERSHIP

Most people spend as much as 70 percent of their time communicating (speaking, listening, writing, reading). Certainly no human skill is more important. Communication has made possible an incredible store of knowledge and is vital to social growth. Because purchasing managers must achieve results through others, they must be skillful communicators. It has been estimated that as much as 80 percent of a manager's time is spent in some form of communication. Obviously then, a goal of the purchasing manager should be to be as effective a communicator as possible.

Communication is a two-way exchange of ideas and information that leads to a common understanding. The word *communication* is derived from a Latin word meaning "common," or "shared by all alike." If one person is unable to achieve a common understanding with another, then communication has not taken place. For example, if the president of an organization sends a letter to all employees explaining a new organizational policy, and the employees fail to read the letter, there has been no communication. If an instructor makes a point in a lecture that is misunderstood by the students, communication has not occurred. Communication takes place only when an idea is transferred from one mind to another in a way that is understood and can be acted on. "Communication is an art form, requiring expertise and continual practice."[4] Figure 8-2 outlines the communication process.[5]

---

[4]Ballew, Arlette C., *"The Need for Interpersonal Skills," NAPM Insights,* August 1994, p. 17.
[5]Ballew, p. 17.

FIGURE 8-2
The Communications Process

Sender

1. Sender has an idea
2. Idea is put into words
6. Sender interprets feedback to make sure the correct message was received

Receiver

3. Receiver hears (or sees) words (or actions.
4. Translates words (or action) into ideas
5. Receiver feeds back an indication that message is understood, or misunderstood.

Feedback

Feedback verifies the message sent from the receiver to the source and is defined as *the signal the receiver gives the sender to show that the message was received and understood or misunderstood.* Its importance to the communications process is often overlooked. Without feedback, the sender has no way of knowing if communication actually took place. Feedback takes many forms other than the spoken word, such as a smile, handshake, glance, kiss, blank look, a teacher testing a class, or a supplier failing to respond to an RFQ.

A purchasing manager can do several things to improve communication through feedback. First, it is necessary to provide a timely means for feedback with any communication. Feedback must be actively pursued and acted upon. An effective way to get instant feedback is through conversation or group meetings with buyers. Real-time feedback is useful since information deteriorates over time. The question, "Do you understand?" is not an especially useful measure of communication because people almost always answer affirmatively whether they understand or not. In addition, understanding an idea and being able to put it into practice are not always synonymous. It may be necessary to ask the receivers questions or to ask the person to demonstrate or explain what has been expressed. For example, one might show a new buyer how to fill out a purchase order and then have the person complete one for review. Feedback does not have to take the same form as the original message. For example, a manager might post a draft memo of an important proposed purchasing department policy and then hold a department meeting to get feedback from the buyers. The meeting would be held to ensure that the message was read, understood, and deliberated.

The second key facet of feedback is to learn to be sensitive to it. Feedback is not always verbal; it can be communicated in subtle ways. It is necessary to observe the actions of those with whom one is communicating. This includes noticing and interpreting facial expressions, tones of voice, the looks in people's eyes, and the movements of their bodies. Overall, it is important to study very carefully what people are saying and what they are not saying.

Feedback must also relate a precise behavior ("When you conducted those negotiations last week. . . .") and should not be a representation of the actions or analysis of the person's overall character or motivations. Negative feedback should be handled cautiously, but when used appropriately, it can be very effective for modifying behavior. It is more effective when requested and least effective when it is delivered at an inappropriate time or in an inappropriate situation.

> Negative feedback should describe something that the person can do something about. Describe the effect of the behavior on you ("I felt embarrassed" or "I felt proud to have you on our team"). If appropriate, state what you would like to see happen (". . . and I'd like to see your reports before meetings" or ". . . and I'd like you to create our charts from now on"). Then stop talking unless the person asks for clarification or further discussion. Do not pressure the person to respond.[6]

With regard to feedback, it is also important to know that there are two general directions in which formal information can flow within an organization: (1) vertically (up and down the organization); and (2) horizontally (back and forth on the same organizational level). Figure 8-3 provides a diagram of this flow of information. The effectiveness of communication varies widely with the direction of the flow, and the feedback signals will vary accordingly.

Studies of horizontal communication have shown that it can be 80–90 percent effective. This high effectiveness results when people working on the same level have a good understanding of each other's work and problems and anticipate much of what is being communicated. Vertical communication is much less effective than horizontal. Studies have shown that only 20–25 percent of the information leaving a president's office is received and understood by a worker. It is an incredible commentary on organizational communication that the people doing the work receive only 20 percent of the information sent to them from the top.

---

[6]For more information on effective business communications, please see Northey, Margot, *Impact: A Guide to Business Communication,* Third Edition, Scarborough, Ontario: Prentice-Hall Canada, 1993.

FIGURE 8-3
Vertical and Horizontal Information Flow

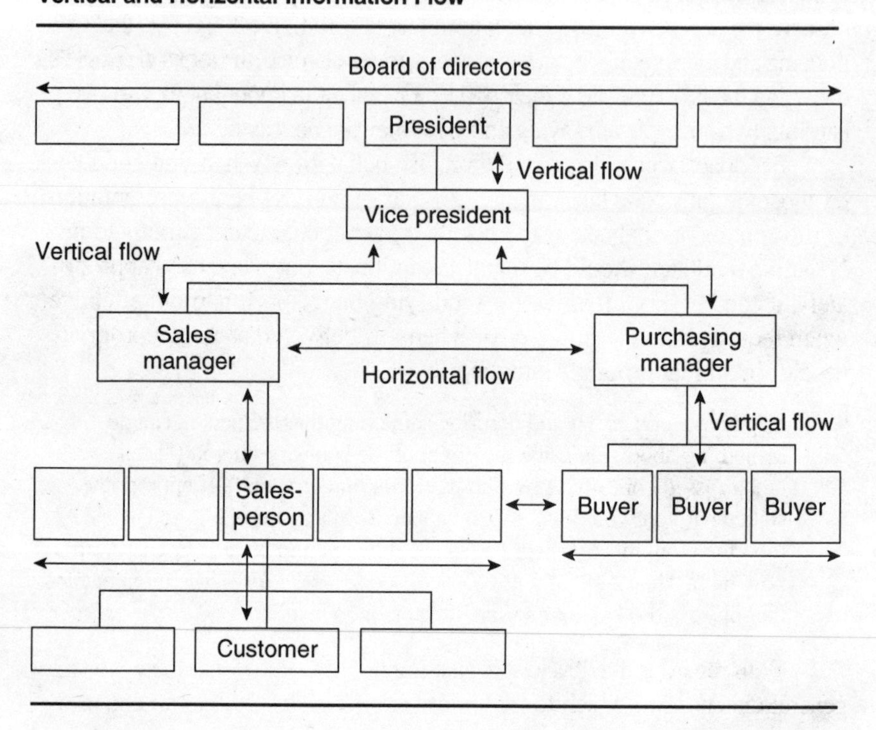

In other words, four times out of every five, they either receive no information or the wrong information. While downward communication is ineffective, upward communication is even worse. Studies have shown that less than 10 percent of what the worker sends to the top is correctly received.

Clearly, there is a huge communication gap between the workers and top management. What can the purchasing manager do to improve communication? The section that follows examines more fully the causes of communication failure and explains in detail how to make your communications more effective.

OVERCOMING BARRIERS TO COMMUNICATION

The major causes of poor communication are (1) the idea that communication is unimportant, (2) a poor mental set, (3) faulty message construction, (4) faulty memory, and (5) failure to establish a means for feedback.

Failing to Realize that Communication is Important

Management too often takes the position that workers and even middle and lower management do not need to know what is happening. Top management thinks that those down the line should do what they are told and not ask questions. Even some middle managers think this way about their first-line managers and workers. Yet studies have shown that employees rank "being in on things" as second or third out of a list of 10 morale factors that influence their work. When management is asked to rank how employees feel about the same 10 factors, it ranks "being in on things" at the bottom of the list.

The mind does not like unanswered questions. If those in the know do not supply the answers, workers take what information they do have and invent answers where information is lacking. To be a good purchasing manager it is necessary to seek the answers to important questions and to quickly pass this information along to others in the organization.

The Mental Set and Faulty Communication

A mental set is one's attitude and is based on one's entire past experience. For example, experiments have shown that poor children think of a half-dollar as being much larger in circumference than rich children do. Five pitfalls of the mental set include (1) stereotypes, (2) fixed beliefs, (3) poor attitude, (4) lack of attention and interest, and (5) lack of facts.

1. ***Stereotypes***—affect communication in two ways: (a) a sender who has stereotyped ideas is likely to distort the message he or she is trying to relay and (b) a receiver who has stereotyped ideas distorts the message when he or she receives it.

   To combat the problem of stereotyping, it is necessary to examine the message for possible stereotypes, to determine what the receiver's stereotyped ideas might be, and to construct the message in a way that will overcome this problem. To be a better receiver, one needs to avoid stereotypical thinking. This is not an easy thing to accomplish, but constant watchfulness will result in tremendous self-improvement.

2. ***Fixed beliefs***—One's beliefs detract from effective communication. Individuals constantly search for things to support what they believe to be true and quickly reject ideas that are inconsistent

with their beliefs. As a receiver, one must be careful not to reject new and different ideas just because they are new and different.

3. ***Poor Attitude***—The attitudes or opinions of the sender and receiver toward each other are very important to communication. Where antagonism exists, communication is unlikely to be effective. However, when the sender and receiver each have a high regard for the other, communication is aided and even the most difficult problem is on the way to being solved.

Part of the task of a manager is to create and maintain a working relationship with employees that promotes an attitude of mutual respect and understanding.

4. ***Lack of attention and interest***—Getting the attention of the audience is the first requirement. With the advent of radio and television commercials, people have become skilled at tuning out things that do not immediately interest them, thus making communication more difficult.

Having garnered the attention of the audience, an effective communicator must turn that attention into real interest by expressly ensuring that the receiver can visualize how the message relates to him or her. There are two basic ways to arouse such interest. The first is through positive stimulation (i.e., show them how to get what they want). The second is through negative stimulation (i.e., show them how to avoid things they do not want).

5. ***Lack of facts***—The last element in a poor mental set is the lack of factual material and/or the misinterpretation of facts. Aristotle pondered the fact that birds in Athens were a different color in the winter than in the summer. He concluded that birds change color. Apparently he did not consider the possibility that birds migrate. Because he misinterpreted the meaning of bird coloring, any messages that he either sent or received about bird coloring were likely to be wrong.

The lack of facts can lead to a failure to communicate. Suppose that Bill Barnes, heading down the corridor toward the coffee machine, overhears the purchasing manager on the phone with the human resources manager. His manager is saying, "You are right,

Martha. Something is going to have to be done about Barnes." Bill is worried and concludes that he must be in trouble. He even thinks he may be fired. Actually, what the human resources manager said was, "Something needs to be done to bring Barnes' pay up to the local market scale." The manager replied, "You are right, Martha, something is going to have to be done about Barnes."

People do not like to have incomplete information, so they take what facts are available and fill in their own idea of what the real facts might be. Sometimes they are right, and they reach the correct conclusion. But just as often they are wrong. A lack of facts in most communication is the fault of the sender. Sometimes the sender unknowingly leaves out important facts. More often, however, important facts are omitted because the sender assumes that the receiver either already knows these facts or does not need to know them. It is important to continually check the messages, not only to see that the facts are correct but also to ensure that enough facts have been included.

Faulty Message Construction

This factor can be a major barrier to communication. Because of faulty message construction, many messages are never received correctly; the receiver is unable to figure out what the message means. Five of the more common faults that can be avoided include:

1. ***Failure to choose the right words.*** In other words, this shortcoming is a matter of semantics. Semantics is the study of the relationship between symbols and what they mean. Words are symbols that represent things, people, and ideas that we want to communicate. But the same word may have 10 to 20 meanings assigned to it. Some of the meanings are totally different, and the only way to express the intended meaning is to carefully place the words in each sentence and to be alert to the emphasis placed on them, whether written or oral.
2. ***Failure to organize the message properly.*** The message should lead the receiver (a) from attention to interest, (b) from interest to main points, (c) from main points to objections and questions, (d) from objections and questions to summary, and (e) from summary to a call for action. Sometimes a summary at the very beginning helps draw the receiver's attention to what follows, but it still does not preclude the need to summarize later.

3. ***Failure to assess the receiver's ability properly.*** Feedback should reveal whether the construction of the message is appropriate. It is often helpful to keep the message as simple as possible. No matter how intelligent or sophisticated the receivers, simple messages have a much greater chance of being received than complicated ones.

Why issue the following. . .

Pursuant to the new policy regarding individual responsibility for punctuality, to be instituted in this department on October 26, time clocks will no longer be utilized as a means for regulating employee working hours. Instead, each employee will be responsible for keeping an accurate written record of actual hours worked, said record to be submitted to the department head prior to the employee's departure each Friday. Forms to facilitate the new procedure will be distributed on October 26.

when it is possible to say. . .

Good news! Nobody has to punch a time clock any more! Starting next Monday, October 26, everyone will be responsible for keeping track of his or her own hours. Every day, write down the hours you actually work. Then give me the record for the week before you go home on Friday. To make the new system easier for everyone, I will distribute forms next Monday. If you have any questions, just ask me.

The second version is a little shorter, a lot clearer, and sets up a system for feedback. Notice that the sentences are, in general, no more than 14 to 17 words long. If a message is very long it may be helpful to repeat the important points using different words.

4. ***Failure to create credibility***. Unless the receiver believes in what is said, the message cannot possibly be effective. And the credibility of one message will depend on the credibility of all the other messages that have been sent. In fact, an individual's credibility is on trial every time a message is sent. A purchasing manager could send 100 messages, but if that purchasing manager is untruthful or has the wrong facts in just one message, his or her competence could be undermined.
5. ***Failure to call for action.*** Managers are often not forceful enough. They send a message but stop short of asking the receiver to take specific action. Hoping the receiver will figure

out what is wanted or assuming that employees know what is wanted is a common mistake. A good purchasing manager makes clear what action is desired and when it is expected.

Poor Memory and Communications

Psychological studies on the problem of forgetting have revealed that the average person forgets more in the first 24 hours after being exposed to new facts than in the next two years. The amount forgotten in that first 24 hours is 70–75 percent. In other words, 24 hours after learning something, individuals typically retain only 25–30 percent of that to which they were exposed. This anomaly explains why a student who neglects to study all term, then hits the books hard the night before an exam, can sometimes score well; there is less time to forget from learning to testing. This process is not recommended, however, because it does not lead to good long-term results. For best results in the long run, a student would do better with several short review periods between learning and testing.

One way a manager can use these findings to become a better communicator is to review important information with his or her employees. A purchasing manager might, for example, hold a departmental meeting to explain a new inventory reporting procedure. At the end of the meeting, it is advisable for the manager to distribute written instructions to the employees and ask them to study the procedure. Posting copies of the instructions on the department bulletin board or using E-mail would also help reiterate the intended message. A couple of days later, another review meeting might be held to provide an opportunity to note common mistakes that are being made. In addition, it is advisable to work with employees on an individual basis.

Another way to improve remembering is to use visual aids. Not only do visual aids get attention and interest, but they also improve reception because the message reaches the brain through both eye and ear. These suggestions require extra work but are necessary for achieving good results. The payoff (if the message is worthwhile) is well worth the extra effort. The quality of a department's work will improve with better communication.

Failing to Establish a Means for Feedback

Although feedback is discussed earlier in this chapter, it is also mentioned here because its absence is a serious barrier to communication. A means for constant feedback is required for good communication. The most important factor in setting up constant feedback is to establish rapport. *Rapport* is a French word meaning "close or sympathetic relationship." When effective rapport is established, feedback often takes care of itself. And rapport is also closely related to good morale. In fact, the combination of good morale among workers and rapport between workers and their managers often leads to effective teamwork.[7]

DEVELOPING TEAMWORK

Teamwork means that everyone in the purchasing department cooperates with each other to achieve the organization's goals. Formal leaders of the group are responsible for establishing and maintaining teamwork, but they must be sincere in applying the aforementioned techniques of effective leadership; workers can easily spot a phony. Another key to success is consistency. A stop-and-start approach will not work. It is not possible to embark on periodic teamwork campaigns. Teamwork is something that has to be aggressively pursued every day. But in the final analysis, the natural results of good leadership and good communication—improved morale and rapport—will do more than anything else to develop teamwork.

KEY POINTS

1. Directing is the use of communication and leadership to guide the performance of one's subordinates toward the achievement of the organization's plans. Directing is most effective if the goals of the organization and the employee coincide.
2. The two basic principles of directing are direction, or a common goal, and unity of purpose, or a desire to achieve that goal.
3. Leadership is the ability to inspire employees toward a goal. Leadership styles range from the outdated oppressive autocrat and the benevolent autocrat to the consultative leader and participatory leader. Influences include the ability and personality of the leader, the ability and personality of the employees, the

---

[7]For more information on effective business communications, please see Northey, Margot, *Impact: A Guide to Business Communication,* Third Edition, Scarborough, Ontario: Prentice-Hall Canada, 1993.

work situation, the style of leadership used by managers above the purchasing manager, and organizational policies and practices.

4. When developing a leadership style, a manager should be consistent, focus on long-term goals, and be oriented toward the needs of the employees as well as the needs of the organization.
5. Effective, positive long-term leadership leads to higher employee morale and productivity.
6. Employees are generally more comfortable with a manager who has a consistent but flexible style, allows some input on decisions that directly affect them, and is employee-oriented, not just task-oriented.
7. Good leadership requires effective communication. A good indication of the effectiveness of communication is drawn from feedback, and it is essential to always provide a forum for feedback and to be attentive to nonverbal types of feedback.
8. Communication and feedback can flow up and down the corporate hierarchy, but horizontal communication between workers at the same level is far more effective.
9. Communicating becomes ineffective when it is not considered important by its intended receivers or when the sender does not consider the mental set of the receiver. The sender must get the receiver's attention, then provide enough facts to maintain the receiver's interest.
10. In forming a message, the sender should write at a level best understood by the receiver. Messages should be carefully worded, organized, to the point, and credible, and must clearly state what actions are desired.
11. People have a tendency to forget information over time. A manager should review and reiterate important information every couple of days. Visual aids or messages also help to increase retention.
12. Overall, teamwork can develop from the consistent efforts of the manager working to establish and maintain rapport with workers, who generally respond favorably to a positive and communicative leadership style.

CHAPTER 9

MOTIVATION

MOTIVATION

Motivation is the internal process that causes a person to initiate behavior to satisfy the work assigned to them in a manner that meets or surpasses expected standards of performance.[1] If there is any one key to "getting work done through others," it is the ability to put other people into motion, in the right direction, day after day. This chapter introduces some of the better known research and thinking on motivation. After describing the contributions of several people, the text describes how the purchasing manager can apply some of their findings.

Taylor's Piece-Rate System

In 1911, Frederick Taylor, father of scientific management, (1) advocated high wages and low cost as the foundation of the best management, (2) described the general principles that render it possible to maintain these conditions (even under the most trying circumstances), and (3) demonstrated the various steps to be taken in changing from a poor system to a better type of management.[2]

To achieve higher wages and low cost, he proposed a piece-rate system as a way of motivating first-class workers:

(a) that each workman should be given as far as possible the higher grade of work for which his ability and physique fit him

---

[1]Badawy, M.K., *Developing Managerial Skills in Engineers and Scientists: Succeeding as a Technical Manager,* New York: Van Nostrand Reinhold, 1982, p. 9.

[2]Taylor, Frederick W., *Scientific Management,* New York: Harper & Row, 1947, p. 22.

(b) that each workman should be called upon to turn out the maximum amount of work which a first-rate man of his class can do and thrive

(c) that each workman, when he works at the best pace of a first-class man, should be paid from 30 percent to 100 percent according to the nature of the work which he does, beyond the average of his class[3]

"Soldiering"

Frederick Taylor realized that a major obstacle to increased production by any kind of incentive was the tendency for workers to *soldier*. He divided soldiering into two categories.

> This loafing or soldiering proceeds from two causes. First, from the natural instinct and tendency of men to take it easy, which may be called *natural soldiering*. Second, from more intricate second thought and reasoning caused by their relations with other men, which may be called *systematic soldiering*.[4]

> The natural laziness of men is serious, but by far the greatest evil from which both workmen and employers are suffering is the *systematic soldiering* which is almost universal under all of the ordinary schemes of management and which results from a careful study on the part of the workmen of what they think will promote their best interests.[5]

> The feeling of antagonism under the ordinary piecework system becomes in many cases so marked on the part of the men that any proposition made by their employers, however reasonable, is looked upon with suspicion. Soldiering becomes such a fixed habit that men will frequently take pains to restrict the product of machines which they are running when even a large increase in output would involve no more work on their part.[6]

Motion and Time Standards

Taylor proposed to overcome the natural tendency of workers to soldier by first setting standards. He used two types of standards: (1) motion standards,

---

[3]Taylor, pp. 28-29.

[4]Taylor, p. 30.

[5]Taylor, p. 32.

[6]Taylor, p. 35.

which define the most efficient procedures for performing a task; and (2) time standards, which determine the amount of time required to perform the task. These standards are used to decide accurately the amount of work each worker should be able to do each day.

> It is not only practicable, but comparatively easy to obtain, through a systematic and scientific time study, exact information as to how much of any given kind of work either a first-class or an average man can do in a day.[7]

The second part of the Taylor plan was to reward (by means of a bonus paid for each piece above the standard) workers who produced an above-average rate of output. Taylor's system was tested at Bethlehem Steel Company at South Bethlehem, Pennsylvania. Tests started in the spring of 1899. Taylor's fear of opposition to the system proved to be well founded:

> Being the first piece work started in the works, it excited considerable opposition, both on the part of the workmen and several of the leading men in town, their opposition being based mainly on the old fallacy that if piece work proved successful, a great many men would be thrown out of work. . . . One after another of the new men who were started singly on this job were either persuaded or intimidated into giving it up.[8]

> In the meantime, however, the first man who started on the work earned steadily $1.85 per day, and this object lesson gradually wore out the concerted opposition, which ceased rather suddenly after about two months.[9]

> By the end of the first year, piece workers were handling 3.5 times as many tons per day as the day workers, and their average earnings were $1.88 per day, while day workers earned $1.15 per day.[10]

In the same text, Taylor cited several other cases where changes in methods and working conditions and the institution of standards resulted in significant increases in output. The question naturally arises, if Taylor's approach to increasing production was so good, why are more workers not paid on a piece-rate basis today? Unfortunately, the piece rate and other monetary incentives, such as profit sharing, cost-savings sharing, and commissions, have not been a panacea.

---

[7]Taylor, p. 45.
[8]Taylor, p. 51.
[9]Taylor, p. 51.
[10]Taylor, p. 54.

Applying Taylor's Philosophies

Taylor, like many managers before and since, thought that money could be used to motivate most workers, but the fact is that money does not motivate most people most of the time. In the great majority of cases, a good pay plan serves only to make workers want to stay with that employer, which is, of course, beneficial to the employer in reducing employee turnover.

One way to use money as a motivator to increase output is to hire only those workers who indicate strongly that money motivates them. One (nonunion) company, Lincoln Electric, has very successfully used compensation as a motivator by addressing it explicitly during hiring practices and by subsequently using a very large bonus system—up to 100 percent of the annual earnings. In a recent series of interviews, several of Lincoln's employees who had been there for a long time indicated very high praise for the company and its pay plan. On the other hand, employees who quit could not say enough bad things about the company, the other employees, and the pay plan. One disgruntled former employee called it a "company of rate busters." In short, they hated everything about their experience at that company.

There are two lessons to be learned about money as a motivator. First, money will motivate perhaps 10–30 percent of the average population to produce more, but it will not have this effect on the other 70–90 percent.[11] Second, money can be used as a motivator (1) if the organization is very selective in its hiring practices, (2) if the monetary incentive is substantial (30–100 percent above base pay, according to Taylor), and (3) if social conditions are at least tolerable. That is, at least some additional needs must be met. Money alone is probably not sufficient motivation.

Hawthorne Studies

Professor Elton Mayo, F. J. Roethlisberger, and their colleagues conducted experiments in behavioral science at Western Electric Company's Hawthorne plant from 1927 to 1932. These famous experiments are usually referred to as the "Hawthorne studies."[12] The Hawthorne plant, which was in Cicero, Illinois, employed 29,000 workers to produce telephone components and

---

[11]The exact proportion has not been scientifically determined. See Whyte, W.F., *Money and Motivation,* New York: Harper & Row, 1955, for a thorough discussion of money as a motivator.

[12]The findings of the studies are described in, Roethlisberger, F., and W. J. Dickinson, *Management and the Worker,* Cambridge, MA: Harvard University Press, 1939.

equipment. Originally starting with a group of only five employees, the study eventually included about 20,000 workers.

The Illumination Study

The first study was designed to determine the effects of lighting on quality and quantity of production. During the period of two-and-one-half years, numerous workers were studied while illumination was reduced from 70 foot-candles to 0.06 foot-candles (0.06 is about the amount of light present on a moonlit night). Even when the light was reduced to this low level, the workers were able to maintain their usual output.

The test failed to furnish conclusive proof that there is a connection between light level and output. It seemed to the researchers that human factors were more important than illumination. Since the workers selected for the test felt that they were receiving special recognition, they seemed to compensate for the less desirable environmental conditions by working harder. The test results led the researchers to investigate the relationship between human relations and output.

The Relay Assembly Test

The researchers identified the assembly of telephone relays, which consisted of putting together 35 parts, for subsequent investigation. An experienced worker could complete the assembly in about one minute, and each operator completed about 500 units per day. Five women were chosen for the experiment. The standard work week consisted of six 9-hour days with no rest periods or breaks during the day except lunch. In the experiment, from two to six 5-minute rest pauses were introduced each day. The outcome of the study was that production rose steadily. The researchers attributed the increase in production to two factors: (1) changes in the working conditions; and (2) supervisors' efforts to gain the workers' cooperation and loyalty.

The next series of tests was to determine what effect a shorter workday and/or workweek would have on output. The experimenters were amazed that output maintained an upward trend during the two-year test period regardless of changes made in the length of the workweek or workday. The best explanation seemed to be that the techniques of supervision had improved, and that this enhancement in supervision rather than the shorter workweek accounted for increased production.

Is Piece-Rate a Good Incentive?

The effect of incentive pay on workers' output was the next area of investigation to be undertaken at Hawthorne. The studies indicated that production increased when piece-rate payment was introduced, but that the increases were due not only to greater pay but to other reasons as well.

> At least two conclusions seemed to be warranted from the testroom experiments so far: (1) there was absolutely no evidence in favor of the hypothesis that the continuous increase in output in the Relay Assembly Testroom during the first two years could be attributed to the wage incentive factor alone; and (2) the efficiency of wage incentive was so dependent on its relation to other factors that it was impossible to consider it as a thing in itself having an independent effect on the individual. Only in connection with the interpersonal relations at work and the personal situation outside of work, to mention two important variables, could its effect on output be determined.[13]

In summary, the Hawthorne researchers concluded that it is a mistake to seize upon one particular thing and decide that the workers can and will be motivated by this factor independently of all others. In addition, they determined that people are more likely to be motivated by social factors, such as what their fellow workers think of them, or how the manager treats them, or what they think of the company, than they are by incentive pay.

Applying Hawthorne

There are two significant conclusions to be drawn from the Hawthorne studies. The first is that human beings are complex animals and can seldom be continually motivated by any one factor. Motivation or lack of it is more likely to be the result of a combination of factors acting upon an individual at about the same time. One factor operates at a time, but different factors have an effect over a period of time.

The second lesson is that the average worker is more likely to be motivated by social factors than by monetary factors. So a good purchasing manager must attempt to create a favorable social environment. This does not imply turning control of the organization over to the employees, nor does it

---

[13]Roethlisberger F.J., and William J. Dickinson, p. 160. Reprinted by permission of the publishers from *Management and the Worker: An Account of a Research Program Conducted by the Western Electric Company, Hawthorne Works, Chicago,* Cambridge, Mass.: Harvard University Press, Copyright ©1939, 167 by the President and Fellows of Harvard College, p. 160.

FIGURE 9-1
Maslow's Hierarchy of Needs

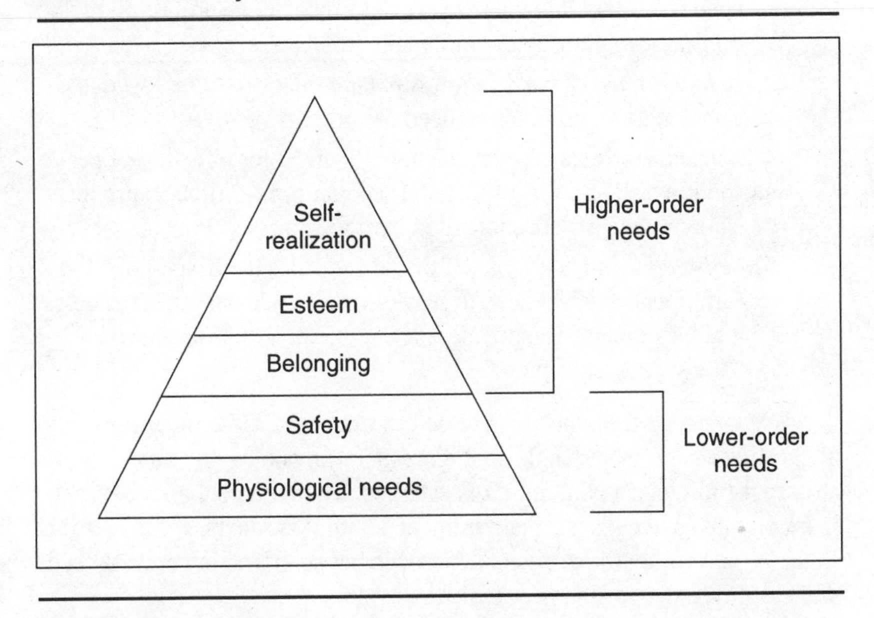

Source: Hierarchy of Needs data from *Motivation and Personality,* Third Edition by Abraham H. Maslow. Revised by Robert Frager, James Fadiman, Cynthia McReynolds, and Ruth Cox. Copyright 1954, © 1987 by Harper & Row, Publishers, Inc. Copyright © 1970 by Abraham H. Maslow. Reprinted by permission of HarperCollins Publishers, Inc.

mean failing to correct employees when they need it. The proper environment can be created by practicing the principles of management found in this text. Employees do respond in a positive way to good leadership.

Maslow's System of Need Priorities

Abraham H. Maslow is famous for classifying human goals and ranking them in order of importance (Figure 9-1). This hierarchy of needs attempts to explain the internal and external forces that motivate a person. The five groups of needs that Maslow defines are as follows:

1. *Physiological needs* are the biological needs for such particulars as food and drink, shelter, warmth, sleep, elimination, and sex.
2. *Safety needs* (or security) include both physical and psychological elements. A person needs to be physically free from illness

or injury, and an individual has a psychological need to feel secure and to maintain the gains that have been made. It is the need for safety that causes people to buy fire insurance on their homes or to put money into banks for safekeeping.

3. *Belonging needs* (or relationships) are the need to be loved and to belong to a group—the need for other people.
4. *Esteem needs* (or respect) include the need for a feeling of personal worth and for respect and recognition from one's group. Many managers are motivated by a need for esteem.
5. *Self-realization* need (or self-actualization) is the desire for self-fulfillment. It is the need to play out a role without regard to how it affects others. The popular phrase that describes this need is "doing your own thing."[14]

Whereas all these needs may be present at the same time, the group at the bottom of the triangle in Figure 9-1 (physiological needs) is the strongest (dominant) goal until it is satisfied. Then the goal above it in the hierarchy takes over as the main influence until it is satisfied, and so forth. Later in this chapter the explicit relationship between the hierarchy of needs and employee motivation is described.

The motivation process, illustrated in Figure 9-2, is one of each person receiving and prioritizing the many stimuli he or she receives moment to moment, and then deciding what course of action or behavior to engage in to try and satisfy the most immediate need. In fact, the word *motive* is from the Old French *motif,* "causing to move," or from Late Latin *motus,* "to move." So motivation is the *internal* drive or cause that puts a person into motion to satisfy his or her needs. While stimuli can be both external, such as heat, light, and sound, or internal, such as hunger, thirst, and sleepiness, motivation can only be internal. This is because it determines what an individual decides to do about a particular stimulus. Therefore, one can *never* motivate someone else. People can only motivate themselves. At best, managers can try to provide positive stimuli that will help employees motivate themselves in the way the purchasing manager wants them to be motivated. This point should not be overlooked as semantic or pedantic. In short, it is the job of management to enhance existing employee motivation, not to create it. How to provide positive stimuli is discussed later in this chapter.

---

[14]Hierarchy of Needs data from *Motivation and Personality,* Third Edition by Abraham H. Maslow. Revised by Robert Frager, James Fadiman, Cynthia McReynolds, and Ruth Cox. Copyright 1954, © 1987 by Harper & Row, Publishers, Inc. Copyright © 1970 by Abraham H. Maslow. Reprinted by permission of HarperCollins Publishers, Inc.

FIGURE 9-2
The Motivation Process

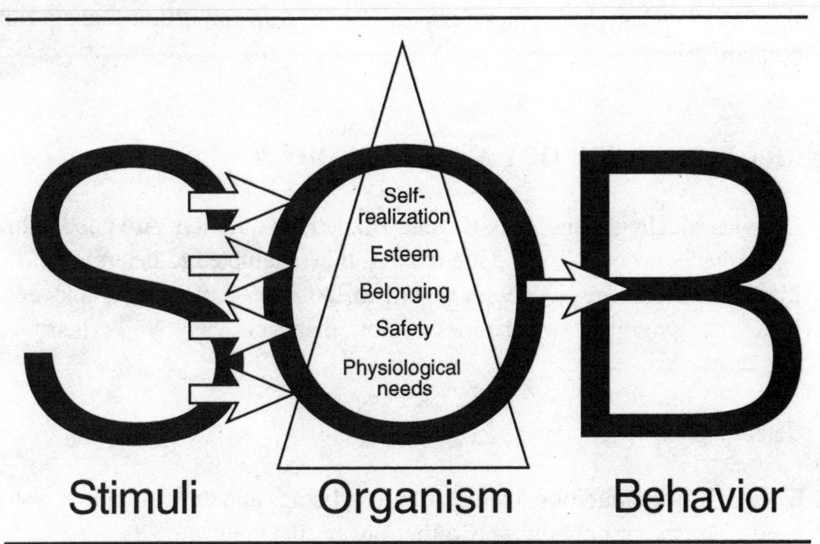

Source: Hierarchy of Needs data from *Motivation and Personality,* Third Edition by Abraham H. Maslow. Revised by Robert Frager, James Fadiman, Cynthia McReynolds, and Ruth Cox. Copyright 1954, © 1987 by Harper & Row, Publishers, Inc. Copyright © 1970 by Abraham H. Maslow. Reprinted by permission of HarperCollins Publishers, Inc.

Applying Maslow

Individuals typically start with the lowermost need of Maslow's hierarchy and concentrate on it until it is satisfied or nearly satisfied. Then the person's attention shifts to the next need in the hierarchy. To help people be motivated, it is necessary for the manager to determine which need they are concentrating on and help them satisfy that need. For example, if an employee is concentrating on being promoted from junior buyer to senior buyer (esteem or self-realization need), the manager may be in a position to provide a stimulus to the employee by saying, "Look, if you do a good job on this project, I'll recommend a promotion for you when the next opening comes along."

One problem that continually arises with this approach is that people's needs change. While in one instance an individual may be trying to fulfill the self-realization need, he or she may soon be concentrating on another need. For example, the employee who is concentrating on self-realization in the morning may in the afternoon be emphasizing safety and security because of a rumor that there is going to be a big layoff. The rumor changed the employee's focus from a higher-order need (self-realization) to a lower-order need (safety and security). Being able to help subordinates maintain this focus as they climb through the hierarchy of needs is an important managerial consideration.

McGREGOR: THEORY X AND THEORY Y

Douglas McGregor and Alex Bavelas studied the question: Are good managers made or born? To find the answer, they attempted to determine how managers see themselves with regard to the management of employees. McGregor presented two theories of how managers view work, theory X and theory Y.

Theory X

Theory X is the traditional view of human beings and their attitude toward work. The major underlying assumptions of this view are:

1. The average human being has an inherent dislike of work and will avoid it if he can.
2. Because of this human characteristic of dislike of work, most people must be coerced, controlled, directed, or threatened with punishment to get them to put forth adequate effort toward the achievement of organizational objectives.
3. The average human being prefers to be directed, wishes to avoid responsibility, has relatively little ambition, and wants security above all.[15]

McGregor did not believe these assumptions to be completely false, reasoning that they would not have persisted if there was not considerable evidence to support them.[16] However, he did believe that these assumptions should be revised and corrected to conform with the findings of modern-day behavioral research.

---

[15]From McGregor, Douglas, *Human Side of Enterprise,* New York: McGraw-Hill, Inc., 1960, pp. 33-34. Reproduced with permission of McGraw-Hill.

[16]McGregor, p. 35.

Theory Y

Theory Y is the modern view of people and their attitude toward work. It is based on knowledge about human behavior and motivation that started with the Hawthorne studies in 1927. This new set of assumptions based on modern research, as stated by McGregor, includes the following:

1. The expenditure of physical and mental effort in work is as natural as play or rest.
2. External control and the threat of punishment are not the only means for bringing about effort toward organizational objectives. Man will exercise self-direction and self-control in the service of objectives to which he is committed.
3. Commitment to objectives is a function of the rewards associated with their achievement.
4. The average human being learns, under proper conditions, not only to accept but to seek responsibility.
5. The capacity to exercise a relatively high degree of imagination, ingenuity, and creativity in the solution of organizational problems in widely, not narrowly, distributed in the population.
6. Under the conditions of modern industrial life, the intellectual potentialities of the average human being are only partially utilized.[17]

McGregor characterized the primary differences between the two theories as summarized in Table 9.1.[18] He realized that theory Y assumptions were easy to accept but hard to implement. McGregor also wanted managers to understand that theory Y was only a set of assumptions, not facts, and that there might be other valid theories of management. In other words, theory Y was an imperfect set of assumptions about the nature of human beings and not a management strategy. The pragmatic aspects of adopting theories X and Y are discussed subsequently in this chapter.

---

[17] McGregor, pp. 47-48.

[18] McGregor, adapted from p. 48.

TABLE 9.1
Theory X Versus Theory Y

THEORY X	THEORY Y
1. Static managerial strategy	1. Dynamic managerial strategy
2. Human growth & development are not possible	2. Human growth & development are possible
3. Single, absolute control	3. Need for selective adaptation is recognized
4. Worker participation in the organization is limited by human nature	4. Worker participation in the organization is limited by management's ability to lead
5. Ineffective organizational performance is due to the nature of the human resources with which management must work	5. Inefficient performance is the result of poor management

Source: Douglas McGregor, *Human Side of Enterprise*, New York: McGraw-Hill, Inc., 1960, pp. 33-34. Reproduced with permission of McGraw-Hill.

HERZBERG'S MOTIVATION HYGIENE THEORY

Frederick Herzberg conducted field research based on the idea that satisfying work contributes to good mental health. His two-factor theory is called *motivation hygiene.* Hygiene, of course, is the science of preserving health and, in this case, mental health. His theories on motivation have had a profound impact on management.

TABLE 9.2
Satisfiers

Satisfiers
Achievement
Recognition
Work itself
Responsibility
Advancement & growth

Satisfiers

Herzberg classified factors affecting worker motivation as either satisfiers or dissatisfiers. The factors that were cited as primary causes of satisfaction and motivation were called *satisfiers,* shown in order of importance in Table 9.2. It is important to note, however, that the absence of these factors did not necessarily result in dissatisfaction, as had been suggested in previous one-dimensional studies, but rather in feelings of no satisfaction.

According to Herzberg's theory, satisfying any or all of these six factors is motivational, as is the joy of doing satisfying work. Herzberg held that job satisfaction can best be achieved through job enrichment, which suggests that workers should be allowed to make more decisions, specialize in work they like, advance in the organization, and so on. In other words, the design of the job and the supervision of the employee should make it possible for him or her to achieve the six satisfiers listed above.

Job enrichment should not be confused with job enlargement. Job enrichment makes a job more interesting because it allows for greater individual freedom, while job enlargement is just adding more work. For example, if a buyer who has never bought capital equipment before, but has always wanted to, is assigned some capital equipment buying, we could say his or her job has been enriched. However, if a buyer has always bought bolts and now management adds the buying of nuts, that is enlarging, not enriching. The differ-

TABLE 9.3
Dissatisfiers

Dissatisfiers
Job security
Status
Company policies
Working conditions
Supervision
Personal life
Interpersonal relationships
Salary

ence is that buying nuts would not be much of a challenge or a chance to grow, whereas adding capital equipment would be adding greater responsibilities.

Dissatisfiers

Similarly, Herzberg classified hygiene or environmental factors as dissatisfiers. They are listed in order of importance in Table 9.3. If any of these elements is inadequate or missing, the worker is likely to be dissatisfied with the job. If these factors are adequate, they do not actively motivate the worker to do a good job; they merely prevent dissatisfaction.

Herzberg's theory holds that the satisfaction of these eight factors does not motivate workers. However, if employees are dissatisfied because of one or more of these factors, their motivation is reduced; and—Herzberg maintained—trying to improve motivation when a state of no dissatisfaction exists is futile. Good salaries do not make employees work harder, but poor ones may make them want to take it easy or quit the company. The employee justifies a slowdown by thinking, "Why should I break my back when these creeps are exploiting me?" To summarize, the satisfaction of dissatisfiers does not motivate workers; it helps only to create favorable conditions in which employees can be motivated through the gratification of some or all of the six satisfiers.

Applying Herzberg

Maslow and Herzberg express essentially the same ideas in different ways. Herzberg's job content factors, or satisfiers (achievement, recognition, work itself, responsibility, advancement, and growth), are roughly equivalent to Maslow's higher-order needs (self-realization, esteem and belonging). But environmental factors, or dissatisfiers (job security, status, company policy, working conditions, supervisors, personal life, interpersonal relations, and salary), are all lower-order needs (safety and physiological).

The difference between the two theories appears to be that Maslow sees lower-order needs as motivational, whereas Herzberg does not. For Herzberg, the lower-order needs (dissatisfiers) detract from motivation but are not motivators in themselves. In practice, whether you call them lower-order needs or dissatisfiers does not make too much difference.

To use the two-factor theory in the workplace, purchasing managers should first concentrate on the hygiene factors and try to eliminate situations that cause dissatisfaction (which Herzberg believed to be the more basic of the two dimensions). Subsequently, it is possible to introduce opportunities for achievement, recognition, work enrichment, responsibility, advancement, and growth.[19]

OUCHI'S THEORY Z

In the early 1980s William Ouchi conducted an intensive study of major Japanese firms to identify the management characteristics that contribute significantly to their success. He found one common thread running throughout the total fabric of Japanese management. Virtually all Japanese firms employ

---

[19]Herzberg, Frederick, Bernard Mausner and Barbara Snyderman, *The Motivation to Work*, New York: Wiley, 1959; Used with permission. (Notwithstanding the profound impact of the two-factor theory on management, many organizational behavior researchers have dismissed Herzberg's theory and have since adopted far more complex and valid conceptualizations of motivation.)

TABLE 9.4
Japanese Versus American Model

JAPANESE ORGANIZATION	AMERICAN ORGANIZATION
Lifetime employment	Short-term employment
Slow evaluation & promotion	Rapid evaluation & promotion
Non-specialized career paths	Specialized career paths
Implicit control mechanisms	Explicit control mechanisms
Collective decision making	Individual decision-making
Collective responsibility	Individual responsibility
Holistic concern	Segmented concern

Source: William Ouchi, *Theory Z: How American Business Can Meet the Japanese Challenge,* Reading, MA: Addison-Wesley Publishing Company, 1981, p. 58. Reprinted by permission of the publisher.

a company-wide managerial philosophy built around an overriding concern for the individual—employees and managers alike. The extensive use of quality circles and other techniques of participative management appears to the Western observer to represent, in part, an extension of the theory Y concept. Ouchi calls this Japanese approach theory Z. In addition, theory Z successfully integrates the achievement of personal goals by employees with the collective goals of the firm. Key elements that characterize these comparisons are shown in Table 9.4.

McGregor and Ouchi

The main thing to remember about McGregor's theory Y is that it is a philosophy of management which holds that people are better than most managers think. Given the proper support and leadership, they are more productive than when pushed or coerced into working. Ouchi's theory Z really shows why many Japanese companies have been successful in adopting what McGregor characterized as theory Y and why many American companies have been less successful because they have clung to their theory X ways. Unfortunately, many American managers at these companies probably truly believe they are practicing theory Y, but if one observes their actions or, more important, asks their employees, it is all too apparent that the vestiges of theory X remain.

TOWARD A MORE MOTIVATED PURCHASING DEPARTMENT

To begin this process of achieving a more motivated department, the manager must examine his or her own value system. Without a sound employee-oriented foundation, motivation will not be at its maximum. It is important to make changes where necessary to ensure that policies and practices reflect a theory Y orientation, which determines how people are treated. Assuming that the right balance between being people-oriented and task-oriented exists, the next problem is how to build on these values. The answer often lies with job enrichment.

Physiological and safety needs have been described as low-order needs. Most people who work in purchasing departments should be able to at least satisfy their basic low-order needs with their salaries and benefits. Having addressed these considerations, purchasing managers can focus greater attention on the higher-order needs of belonging, esteem, and self-realization. These are the needs that often are under-satisfied at work but frequently can be enriched.

By taking another look at Herzberg's list of satisfiers it is possible to see how a buyer's job might be enriched:

1. *Achievement:* Buyers might have a higher sense of achievement if they are given the means to tackle large negotiations or solve a supplier delivery or quality problem. Similarly, a clerical person might be trained to do some buying.
2. *Recognition:* This could be anything from a pat on the back to sending a letter to the president of the company telling how the buyer has saved X dollars.

3. *Work itself:* Like achievement, this means expanding the buyers' scope of activities and letting them make more and bigger decisions.
4. *Responsibility:* The purchasing manager may trust them to operate with less supervision (e.g., signing their own purchase orders), for larger amounts, or to make other decisions.
5. *Advancement:* Along with pay increases it may be useful to consider changes in titles such as trainee, junior buyer, buyer, senior buyer, and purchasing agent.
6. *Growth:* This can be accomplished by giving employees opportunities to learn more about different jobs by attending NAPM meetings and conferences, sending them to courses and workshops, operating a lending library, and by rotating jobs.

By trying to enrich people's jobs, as well as by being a good participative leader, managers will develop, over time, a department that is highly motivated. The true essence of leadership is followership; to lead one must first be of service to his or her followers. In addition to motivation, however, purchasing managers are often confronted with formal and informal group relationships that affect daily and longer-term supply management activities.

Quality Leadership[20]

Many authors have criticized North American management methods as being outdated and diminishing the full potential of organizations by failing to capitalize on the inherent capabilities of the work force. Existing organizations that adhere to a chain of command and established hierarchy of objectives, standards, controls, and accountability have adopted a philosophy that is imbued with the vestiges of management by results, whereby employee performance is directed and evaluated according to numerical goals. According to W. Edwards Deming and his followers, as advocated by Joiner Associates, the use of numerical goals to guide and judge performance fosters a host of problems (Table 9.5).

In contrast, these authors propose the concept of quality leadership that emphasizes *results* by examining the underlying *methods*. The essence

---

[20]For a discussion of quality leadership, please see any of the numerous works by Deming, W. Edwards, or Scholtes, Peter R., *The Team Handbook: How to Use Teams to Improve Quality,* Madison, WI: Joiner Associates, 1988.

TABLE 9.5
Problems Associated with Numerical Goals

Problems Associated with Numerical Goals
Short-term thinking
Misguided focus
Internal conflict
Fudging the figures
Greater fear
Blindness to customer concerns

Source: Peter R. Scholtes, *The Team Handbook: How to Use Teams to Improve Quality*, Madison, WI: Joiner Associates, 1988, p. 1-5.

is to arrive at the underlying root causes of problems and preclude them from happening in the future. The intent is to continuously improve work processes so that the final good or service exceeds customer expectations. Simply buying what design or manufacturing requests is replaced with a seamless process that captures the added value inherent in the supply chain. Information flows early and often, supply team members operate in a risk-free environment and share both risk and opportunity, and the end result is a service provided to the customer that delights them by how effectively it satisfies their needs—even those latent demands they had not previously considered. The key principles of quality leadership are shown in Table 9.6.

Empowerment

One of the key outgrowths of quality leadership is the concept of empowerment. Most organizations accept that empowerment works to build people so they will develop and act, often without supervision, to contribute to the organization and themselves. Empowerment includes being empowered (creating

TABLE 9.6
Principles of Quality Leadership

Principles of Quality Leadership
Customer focus
Obsession with quality
Recognizing the structure in work
Freedom through control
Unity of purpose
Looking for faults in systems
Teamwork
Continued education & training

Source: Peter R. Scholtes, *The Team Handbook: How to Use Teams to Improve Quality,* Madison, WI: Joiner Associates, 1988, p. 1-11.

power for oneself) *and* giving power to other groups for organizational benefit. For empowerment to work, there must be a vision and strategy of where the organization needs to go. Managers need to delegate more (i.e., lead rather than manage), and employees must have more say in daily and future work. Implementing empowerment requires serious background work, preparation of all people involved, careful training, and the development of interpersonal skills, incentives, resources, and action plans.[21]

---

[21]Cavinato, Joseph L., "Empowerment: Extending Purchasing's Effectiveness," *NAPM Insights,* December 1992, p. 6.

Rightsizing/Workload

Rightsizing purchasing is the process of shaping and controlling its growth to improve its effectiveness and efficiency. To do this, purchasers must show the benefits of the department to help prevent unjustified cuts. The key is making the purchasing service tangible and evident by reporting cost savings and value creation. Identifying low priority or low value work allows for the consolidation of resources and a focus on higher leveraged activity. Speed and simplicity are pivotal elements. Rightsizing does not always involve moving work out of the organization. It simplifies the process and reduces the length of time to perform a function, thereby reducing its cost.[22]

Motivation and the Supply Team

Ironically, these motivation tenets also apply to members of the supply chain, and individual motivation techniques can often be applied to organizations interested in enhancing the effectiveness of the supply team. In this context, the same principles of team building that lead to higher performance within and between organizations also contribute to learning and change and, ultimately, to sustained competitive advantage. Purchasing professionals, as managers of external manufacturing, must be at the heart of this evolution in providing incentives to members of the supply team.

GROUP DYNAMICS

Formal Work Groups

Formal work groups refer to the manager's relationship with individuals as members of a formal purchasing group (i.e., the department or functional sections within the department).

Informal Work Groups

Within every formal organization, experience demonstrates that one also finds the existence of one or more informal groups. These groups typically are fairly small and are structured informally around specific interest patterns of the

---

[22]Budding, J., "Shaped for Efficiency," *NAPM Insights,* August 1991, pp. 12-13.

members. They may be social groups, special interest groups, or sometimes pressure groups pushing for change. Whatever the case, such informal groups are an integral part of the departmental organization, and their attitudes and actions can either assist or deter the attainment of departmental objectives.

A wise purchasing manager attempts to utilize the potential influence of informal groups in a constructive manner. To do this, he or she must first recognize the existence of such a group and identify the informal group leader(s). Then, by practicing the concepts of open communications and group involvement in the decision-making process, the manager attempts to align the objectives of the informal group with the objectives of the department. This represents an extension of the participative management strategy employed in dealing with individual and formal work groups within the department. Specific approaches which can be utilized are:

1. Through various individual, committee, and brain-storming techniques, solicit appropriate input on decisions that affect individuals and the informal group. To the extent possible, utilize this input in reaching genuine group oriented decisions.
2. Create and develop a work climate and a reward system that encourage teamwork and cooperation.
3. Attempt to develop the degree of informal group cohesiveness that produces a positive influence on the activities of the formal work group.

The purchasing manager's goal in utilizing this integrative approach is to promote cooperation of the various groups—formal and informal—in daily activities that contributes to attainment of the department's overall objectives.

KEY POINTS

1. Taylor's piece rate system used the output of a "first rate man" as a method for setting compensation and work rates for employees.
2. Taylor used motion and time standards to determine the amount of work an employee is capable of. Any performance above this rate is rewarded with bonuses. Unfortunately, monetary incentives work for only about 10–30 percent of the population.
3. The Hawthorne illumination study, originally designed to test effects of lighting conditions, prompted research on the study of the relationship between human relations and output. The relay assembly test indicated a relationship between supervision and production.

4. The Hawthorne studies showed that workers are more likely to be motivated by social issues than incentive pay. Workers must be continually motivated by different factors over time.
5. Maslow defines the needs that motivate a person, starting with the lowermost needs, as physiological needs, safety or security needs, and moved onto higher level needs of belonging, esteem needs, and self-realization. It is the job of management to enhance employee motivation, not to create it.
6. Increased motivation results from an environment in which lower-level needs are fully met and the employee is free to focus on higher-level needs, or satisfiers. These include achievement, recognition, work itself, responsibility, advancement, and growth.
7. A manager must determine upon which needs an employee is concentrating and help him or her satisfy those needs. Managers must help keep employees focused on achieving the higher-order needs.
8. McGregor's theory X indicates that productive work is limited by human nature; his updated theory Y states that productive output is actually a function of management style and quality, not human nature.
9. Herzberg classified factors affecting worker motivation as either satisfiers or dissatisfiers. Satisfiers include achievement, recognition, advancement, and growth in a job. Dissatisfiers, or negative motivators, include factors such as status, working conditions, supervision, salary, and job security.
10. Ouchi's theory Z shows how Japanese companies have been able to adapt the ideas of McGregor's theory Y, in which people are better than managers think. This is shown in the Japanese use of quality circles and participative management styles.
11. To develop a motivated purchasing department, managers must focus attention on their employees' needs of belonging, esteem, and self-realization. Job enrichment is one tool through which these needs can be addressed.
12. Central to the success of any motivational plan is the degree to which the intended motivator addresses the needs of the individuals for whom it was designed.
13. Team building, empowerment, and rightsizing/downsizing are having a profound impact on the purchasing environment.
14. Purchasing professionals must be at the core of efforts to apply motivation concepts within the supply chain.

CHAPTER 10

CONTROLLING AND EVALUATING

When defining control, Henri Fayol, the French father of management, wrote, "Control consists in verifying whether everything occurs in conformity with the plan adopted, the instructions issued, and principles established."[1] In other words, control is the process of establishing standards of performance, and evaluating actual performance against these standards. Controls are used to monitor or validate the actual progress of a plan and to quickly point out deviations that need to be rectified. It is important to emphasize, however, that control does not necessarily imply being "in control." It is not intended to suggest that centralized, hierarchical styles of leadership are needed. In fact, the contrary is probably preferred.

If plans were always right and if things always went according to plan, controlling would be an unnecessary function. But plans are often imperfect, and things do not always transpire as expected. Frequently, Murphy's laws apply: (1) if anything can go wrong, it will; and (2) of all things that cannot go wrong, some will. Effective control enables purchasing managers to measure, evaluate, and correct deviations from the plan, and, in essence, to mitigate Murphy's impact on the bottom line. Figure 10-1 depicts the iterative control process.

Evaluating is the process of determining values or amounts of specific activities. It necessarily entails determining the underlying causes for significant deviations from planned performance and subsequently developing alternatives to preclude such actions from occurring again. Evaluation often includes both qualitative and quantitative appraisal of value. Objective and subjective evaluation in purchasing occurs at several

[1]Fayol, Henri, *General and Industrial Management,* London: Sir Isaac Pitman and Sons, 1969, p.107.

FIGURE 10-1
The Control Process

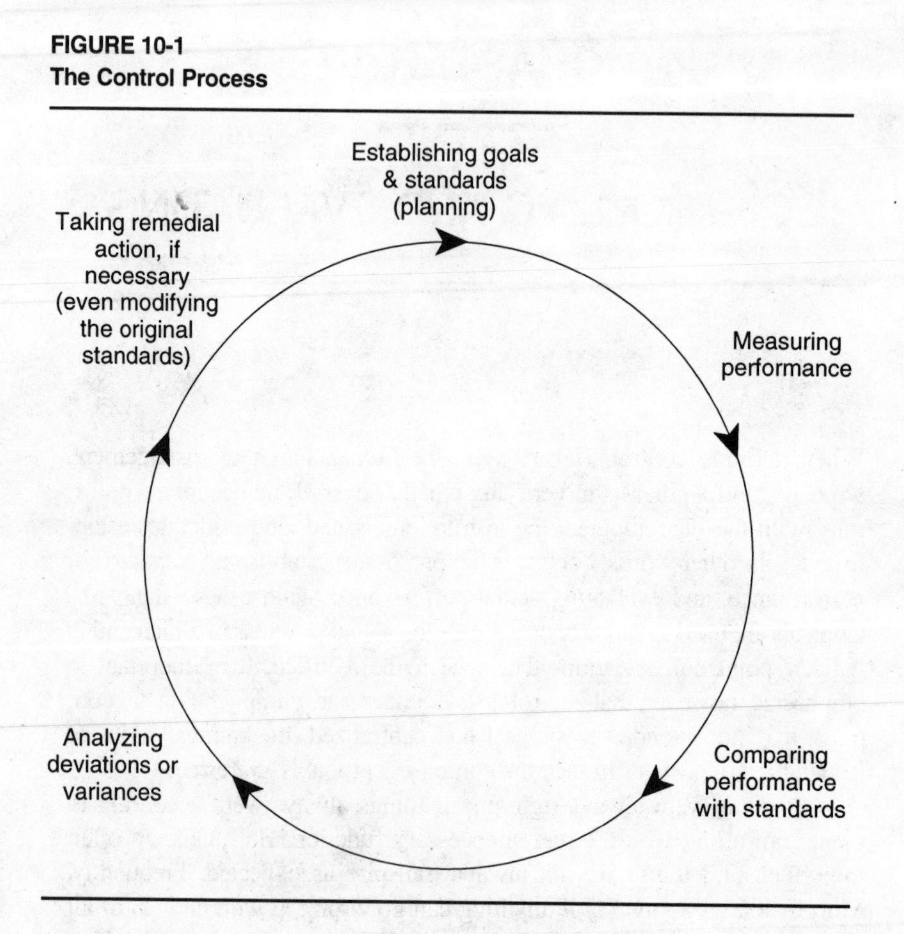

different levels, including the individual, the group or team, the product, service or project, purchasing as a whole, and the management of the supply team. Continuous evaluations compare operating results with established plans and objectives for the department or personnel. Independent or third-party audits are typically made by someone outside the department or the organization.

A couple of years ago, the director of purchasing of a large Midwestern company established a goal to reduce the cost of goods purchased by $3 million that year. It was recommended that he ask each purchasing agent, buyer, and junior buyer to write cost-savings objectives for the year. The director subsequently launched a program to train those involved to write cost-savings objectives. After the training, he directed each individual to prepare a list of objectives for cost savings in purchasing. The objectives totaled a saving of $2.4 million, which the director accepted as a reasonable goal for the department.

A year later when asked if his department was meeting its cost-savings objectives, the director of purchasing said he didn't know! After the training, he had been too busy to check on (control) the progress his subordinates had made toward their objectives. But he added that he felt no one had accomplished the objectives.

What happened in this case unfortunately happens too often in too many organizations. A new program is identified as the panacea for a particular ill. Plans are made, and the program is organized. Those who are to enact the plans are given direction and are perhaps even motivated to achieve the desired outcomes. But the person responsible for the program consequences exercises little or no control over it. The natural result of uncontrolled management is the death of the program because of a severe case of negligence.

What this director of purchasing should have done was to establish measurable standards at the outset and to require monthly progress reports from each person. If the results were encouraging, then the director should have personally acknowledged the progress. If a report suggested that progress was behind the timetable, then the director should have scheduled a meeting with each member of the department and his or her supervisor to ascertain why and what it would take to get this individual's objectives back on schedule. Many more programs and projects of this sort are started than are ever finished. Probably the largest single cause of failure to carry projects through to a successful conclusion is lack of control.

It is better for a purchasing manager to start a small number of special programs and projects and successfully complete them than to start many and complete very few. One suggestion is to adopt the "three theme" principle. In applying this concept, a manager advocates no more than three major themes (e.g., quality, cost control, or value-added activities). These themes are continually reinforced using the "many bullets" principle; i.e., using rewards and recognition, interest level, and any other means of emphasizing them. It is essential, however, to reiterate the importance of these themes and to continually tie daily activities to them. Good purchasing managers are known for the number of programs and projects they complete, not for how many they start.

The need for controlling is based on the fact that things can and do go wrong. The purpose of control is to discover when things are going wrong at an early stage so that the manager can replan, reorganize, redirect, and remotivate in order to overcome any problems.

THE RESPONSIBILITY OF THE PURCHASING MANAGER IN CONTROLLING

The purchasing manager's role can be divided into two main areas:

1. Measuring performance of subordinates according to standards planned by others, such as top management, accounting, purchasing, personnel and so on
2. Measuring performance of subordinates according to standards that managers set when establishing the department's activities

Neglect and Inefficiency Promote High Cost of Care

An example provided by Dean Ammer describes what happens when things get out of control. In the ten years between 1962 and 1972—remembered by hospital administrators as a period in which nurses and other hospital personnel first showed their bargaining muscle and began to win huge raises for themselves—hospital payroll costs increased by 236 percent, from the 1962 level of $13.12 per patient-day to the 1972 level of $44.17 per patient-day. Nonpayroll costs increased during this same period by 349 percent, from $6.41 per patient day to $29.72 per day. Thus, the rate of increase of nonpayroll costs was almost half again greater (349 versus 236) than the rate of increase of payroll costs. The major "villain" as far as hospital costs are concerned was not the hospital employee but the various recipients of nonpayroll costs. And, of course, hospital suppliers collectively are overwhelmingly the major beneficiaries of increases in nonpayroll costs.[2]

---

[2]These and subsequent figures on hospital operating costs are based on estimates in the annual A.H.A. Guide issues of Hospitals. Source: Ammer, Dean S. *Hospital Materials Management: Neglect and Inefficiency Promote High Costs of Care,* Boston: Bureau of Business and Economic Research, Northeastern University, 1975, pp. 1-2. Reprinted with permission.

What to Control

A manager should develop systems to control everything planned and everything that others have planned for which he or she is accountable. The elements described in the previous discussion on planning are applicable in this context. They are as follows:

1. *Objectives*—of the department and organization
2. *Programs*—those being planned or currently under way
3. *Standards*—of both quality and quantity
4. *Policies*—of the department and organization
5. *Procedures*—generally of the department but also of the organization
6. *Methods*—usually departmental ways of doing things
7. *Budgets*—such as sales, expense, capital, and materials budgets

Again, the key to knowing what to control is to look at the specific plan in addition to the responsibilities that have been delegated by others.

THE MANAGEMENT PROCESS: INITIAL PHASE

As described earlier in this text, the management process is often iterative because it has to reflect changes in the environment, priorities, etc. The first phase of the management process for a given task is typically sequential. The cyclical sequence entails planning, organizing, directing, motivating, controlling, and evaluating, which may not always apply to subsequent phases of the same project. The next major section of this text covers the differences in detail. The purpose here is to explain why control must be programmed into the initial phase of the management process from the very beginning and how to implement it.

Planning For Control

During the planning function, managers should think about how they are going to check up on (control) a particular plan. It is a logical, even ideal time for this. When you are making initial plans, develop controls at the same time. During planning, it is necessary to consider: (1) what factors to control, such as cost, time, quality, quantity, and so on; (2) who should do the controlling; (3) how often and when the check-ups should be conducted; and (4) what the best methods of control are.

Organizing for Control

This book describes organizing as providing, in advance, those things needed to carry out a plan. If the control plan calls for using a late delivery report form as a control device, then in the organizing phase the manager should ensure that the late delivery report form is prepared and ready to use at the same time each day. If the control device is a budget (e.g., the purchase of nonferrous metals), it is necessary to organize for control by making sure that those responsible for meeting the budget have a copy of it before the plan is implemented. In addition, the manager should provide a summary of the actual expenses to each person at appropriate intervals so all have the information they need to control the plan.

Directing and Control

Controlling can be effective when a manager expects his or her people to do a good job, conveys what is expected, and lets them know when they succeed. The following are some suggestions to encourage the use of controls by all employees:

1. Establish a mutual interest in working toward the goals. Effective control is virtually impossible unless each person involved knows the immediate goals of the department and works toward meeting those goals. When all employees know what is expected of them, how these expectations were determined, why various standards were developed, and how they will benefit from their accomplishment, they develop an interest in meeting these goals. This step is really a combination of directing and motivating.
2. Communicate effectively to achieve effective controls. Poor performance can often be traced to poor communication. Employee errors might have occurred because employees did not know what they were to do or how they were supposed to do it.
3. Explain the control measures to be adopted and why and how they are to be used.

Control and Motivation

The typical purchasing manager spends much of his or her time controlling—that is, following up on the daily work schedule of each employee. For most purchasing managers, the constant need to oversee each employee's work is distasteful. Most purchasing managers would like to be able to distribute assignments to employees with the expectation that they would achieve the desired outcomes correctly and timely and return for another assignment when they finish. In reality, many managers must continually validate the work of their employees. A second issue is that, after a while, following up on employees day after day is just plain boring. Finally, constant follow-up by supervisors is irritating to the employee.

Remembering that control does not mean "in control," effective managers are not compelled to control too closely; rather, they control at arm's length, ensuring that their subordinates know what is expected of them and what the control methods are. They give their employees a chance to assume responsibility for their own actions, and they say, in effect, "I trust you"—which often helps to motivate the employees.

THE MANAGEMENT PROCESS: SUBSEQUENT PHASES

In practice, there may be a need for only a single phase of the management process. That is, a plan is developed and organized, workers are directed and motivated to execute it, and control shows that the plan is completed as anticipated. This ends the project. However, if the control function shows that activities do not progress as planned, corrective action is necessary. Corrective action is not, strictly speaking, control. Instead, it involves replanning, reorganizing, and redirecting.

If control techniques reveal that the original plan has not succeeded, it is then necessary to revert to one or more of the other functions of management to correct the situation. This process of repeating management activities constitutes Phase 2. Subsequent phases of the management process may also be necessary. Figure 10-2 depicts this process. It is important to note that it may not be necessary to repeat all the steps in the management process during subsequent phases. Instead, as Figure 10-2 reveals, one may need only to redirect subordinates in Phase 3 before again measuring the results. To take a simple example, assume that during control activities it is observed that one of the new buyers is not filling in the

FIGURE 10-2
Management Process Phases

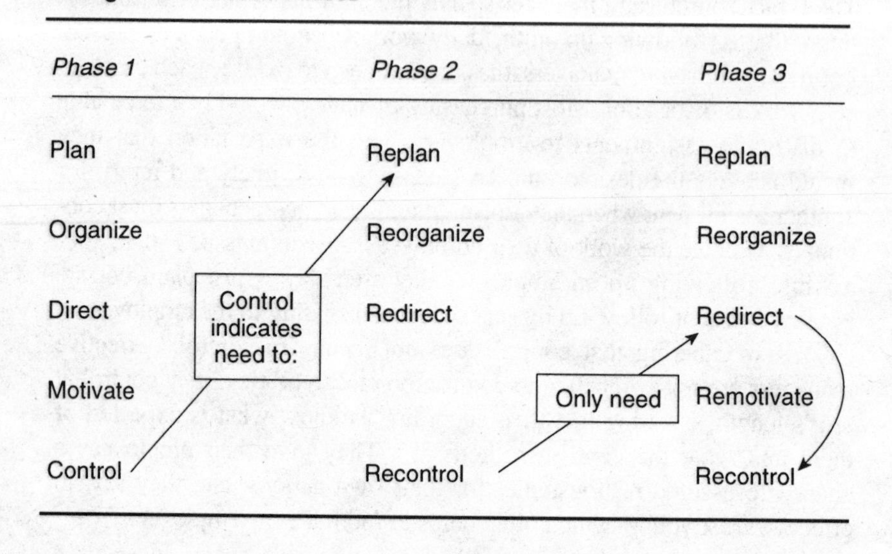

PO correctly. The manager can simply redirect the employee to do it right. This redirection may take the form of training the employee on how to do it right. It is important, however, that the manager not simply rectify the situation himself or herself (i.e., complete a new PO); it is necessary to get to the root cause of the deviation and implement corrective action (i.e., provide the new buyer with an example of a correct PO), which precludes the problem from arising again.

On the other hand, the manager may discover while redirecting the employee that he or she really knows how to complete the PO properly but simply chooses not to do it that way. What happened? In the controlling process it is noted that the requisition is not being properly filled out. The manager concludes from this that the employee is not properly trained, so he or she decides that redirection is the necessary course of action. But in trying to implement the decision, the manager discovers that this conclusion is unfounded; what is really needed is to find out why the employee is not motivated to do it right in the first place.

Or consider the possibility that in trying to redirect the employee to complete the form correctly, it is determined that the form cannot, in most cases, be completed as requested. This new information may lead to replanning, reorganizing, and so forth. The purchasing manager's role in the controlling process must be one of constant follow-up to find out (1) if corrective action is needed, and (2) if so, what method is needed.

Corrective Action

If progress is not proceeding according to plan, the manager must take corrective action. The form of the action depends on what has gone wrong. How soon this must take place is dictated by the nature of the trouble. Some of the chief reasons why corrective action may be needed are as follows:

1. *Poor planning.* Often when managers are planning something with which they have little experience, they tend to be too optimistic. They plan for better and faster results than they can actually achieve. Or they fail to foresee certain problems that will arise. Or they just forge ahead and plan the wrong steps to be taken.
2. *Poor organization.* Sometimes managers fail to arrange for the right circumstances or people needed to carry out the overall plan. This is like going on a picnic and forgetting the hot dogs or receiving 4-inch concrete blocks when 8-inch blocks were ordered.
3. *Poor direction.* Failure in this area is usually caused by poor communication; the managers fail to get the employees to understand exactly what the plan dictates.
4. *Poor motivation.* Employees do not want to carry out the plan.
5. *Changing conditions.* After plans are established and work is under way, conditions may change, thereby creating a need for corrective action to cope with the new conditions. Examples of things that cause changing conditions are (a) machine breakdown, (b) absenteeism, (c) new work orders, (d) materials not being delivered on time, and (e) a strike, fire, or natural disaster (storm or flood).

PURCHASING DEPARTMENTAL EVALUATION

Table 10.1 lists the primary reasons for purchasing departmental evaluation or appraisal.

Meeting Organizational Needs

To evaluate the overall effectiveness of the purchasing operation, one must understand the mission of the total organization and the specific objectives

TABLE 10.1
Purchasing Departmental Evaluation

To determine departmental effectiveness in meeting organizational needs
To determine effectiveness of departmental management
To measure improvement/deterioration
To provide incentives for improvement

of the purchasing operation—to obtain maximum value, prescribed quality, and continuity of supply. Once one understands the organizational mission, this information can be related to the ability to perform tasks which, when combined, ultimately meet goals. Benchmarks are then established to measure the cost/benefit. The purchasing department is expected to carry its weight in the organization.

Departmental Management Effectiveness

The types of factors that the purchasing manager may wish to rate include the capabilities of personnel, the soundness of the organizational structure, the scope of each job, the departmental plans, policies, procedures, and so forth. Such factors influence the potential level of a department's performance and are therefore useful indicators of performance. A list of frequently cited information used to gauge purchasing operations and effectiveness is provided in Table 10.2.

Measuring Improvement/Deterioration

Measurement provides early warning signals of deterioration of performance, which allows managers and buyers to take corrective action when necessary. Measurement also has a built-in incentive; purchasing professionals constantly witness the impact of the competitive environment. Well-established and accepted performance criteria generally provide an objective means of measurement of continuous improvement.

TABLE 10.2
Purchasing Operations and Effectiveness

Purchasing Operations and Effectiveness
Cost reductions resulting from purchase research & value analysis studies
Quality rejection rates for major items
Percentage of on-time deliveries
Number of out-of-stock situations that caused interruption of scheduled production
Number of change orders issued, classified by cause
Number of requisitions received & processed
Number of POs issued
Employee workload & productivity
Transportation costs

Source: Michiel R. Leenders, and Harold E. Fearon, *Purchasing and Materials Management,* Tenth Edition, Homewood, IL: Irwin, 1993, p. 467.

DEPARTMENT-LEVEL EVALUATION

Table 10.3 provides the requisite steps for conducting a department-level appraisal of purchasing.

TABLE 10.3
Purchasing Department Evaluation Steps

Purchasing Department Evaluation Steps
Determine department objectives
Determine appraisal factors & criteria for success
Conduct internal audits/self-governance
Exercise management control in response to results
Compare performance with the plan
Reestablish department objectives

Departmental Objectives

When a manager develops a system to evaluate the performance of any department, a logical starting point is to analyze the departmental objectives. Once this has been done, the organizational structure and the responsibilities assigned to each work group should be examined to determine the impact each operating activity has on the attainment of each departmental objective. This procedure normally discloses the critical activities in the operation where evaluation and subsequent control are most important.

An analysis of departmental objectives made in a purchasing department often reveals that most of the critical points at which evaluation and control should be affected lie within the buying function. A wide range of extremely varied activities is placed under the jurisdiction of each specialized materials buyer. Moreover, examining the basic responsibilities of each buyer discloses that performance in most areas is quite difficult to express in quantitative terms. Thus, the very nature of the purchasing function itself makes it somewhat difficult to establish workload performance standards.

Because of this problem, many organizations have adopted a broad approach to the evaluation and control of their purchasing activities. Most companies attempt to establish performance targets (standards) for the measurable "secondary factors" that contribute to attainment of primary buying objectives. Recognizing that a single factor alone may not provide as accurate an indication of creative buying performance, most companies develop a cross-check by measuring several factors that relate to the same primary objective. For example, buying performance relative to the price objective can be checked from two standpoints: (1) actual prices paid can be compared with target prices; and (2) targets for cost savings resulting from negotiation and from value analysis can be established and actual savings compared with these targets. Thus, two measurements provide a cross-check on attainment of the same primary objective—price. A similar approach can be used in evaluating buying performance relative to each of the other basic objectives.

Evaluation Factors and Criteria for Success

While there are many factors that can be considered when evaluating a purchasing department's performance, typical appraisals consider one or more of the following:

Contributions to profitability—Contribution to profitability is a primary concern in a business. Government agencies and institutions are concerned with total value and with maximizing resource utilization.

Timeliness of actions—A purchasing department's primary responsibility is to support operations. The following can measure how effectively this responsibility is fulfilled:

1. Percentage of overdue orders
2. Percentage of stock-outs caused by late deliveries
3. Number of production stoppages caused by late deliveries

These data can be categorized by material classification, by supplier, or by buyer, depending on the need and purpose of evaluation. It is obvious that a number of different performance factors can be measured to provide a basis for appraising buying proficiency. These factors differ in importance among different firms depending upon the nature of the business and the materials purchased. Most organizations, therefore, do not use all the measures listed for this objective. Each selects those measures that are most useful and most cost effective in its own specific situation.

Prices paid for materials—The following techniques can provide a cross-check on the reasonableness of prices paid for materials:

1. Standard or target prices can be established for major materials. Prices actually paid can then be charted against the target figures to display any significant differences. Another basis of comparison is a materials budget, utilizing standard price data.
2. An organization can develop its own average "price paid" indices for major classes of materials. The trends of such price indices are valuable guides in assessing effectiveness of performance. If developed on a comparable basis, these indices can also be charted against various national commodity price indices published by the Bureau of Labor Statistics and the Department of Commerce. This comparison reveals cases in which a firm's material costs are rising at a greater rate than market prices during an inflationary period.
3. Periodic cost savings figures can be individually charted for savings arising from such activities as negotiation, value analysis, design and materials changes, supplier suggestions, change of supplier, packaging improvements, and transportation cost reduction projects.
4. If an organization engages in forward buying activities, gains and losses from forward buying can periodically be reported to determine forecasting effectiveness.
5. A report of the percentage of purchase orders that are issued without firm prices provides another basis for evaluating and controlling material costs.

The preceding measurements can be classified and sub-classified in various ways to pinpoint responsibility for the problems they reveal.

Material quality—Once material specifications have been established, the most direct measure of quality performance is the percentage, or number, of delivered materials that are rejected by the inspection and operations departments. To check on the improvement of quality specifications, the purchasing manager can also review the value analysis reports dealing with design changes and materials substitutions.

Supplier reliability—The following measurements can be used to indicate the reliability of major suppliers:

1. Percentage of late deliveries and percentage of rejected items, further analyzed and classified by supplier, buyer, etc.
2. Percentage of orders on which incorrect materials were shipped
3. Percentage of orders on which incorrect quantities of materials were shipped
4. Percentage of orders on which split shipments were made
5. Quality and reliability of transportation service offered by various carriers

Suppliers impact a significant part of cost, and their failure to perform must be considered in the search for maximum value. The more reliable the supplier, the less costly it is to deal with them. But supplier evaluation is being transformed by the needs of the marketplace, and purchasing must take a more active role in identifying supply team members based on added value rather than lower cost.

This examination of value transcends the more traditional notions of value analysis and vendor analysis to include concerns for benchmarking. As suggested above, these analyses are necessary but not sufficient. Purchasing must also participate in activities that question basic business processes. What emerges is a fundamental shift away from following work flow and controlling costs to contributing to organizational learning and a fundamental understanding of the firm's core capabilities and skills.

Order quantity and inventory investment—Purchasing's failure to buy the right quantity (i.e., that which keeps the operation functioning, yet minimizes the amount tied up in inventory investment) jeopardizes the optimal cost structure or misuses resources that may be better used elsewhere.

Measurements useful in evaluating how well purchasing invests company funds are:

1. A chart showing target and actual inventory levels in the aggregate and by major material classifications

   This chart is most useful when supplemented with a chart showing inventory turnover rates for the same material classifications. When analyzed together, these charts point out imbalances between inventory carrying costs and material acquisition costs

2. A report of "dead stock" materials carried in stores, resulting from overbuying
3. The number of stock-outs and production stoppages attributed to under buying
4. A list of supplier stocking arrangements that have been negotiated, along with an estimate of resulting inventory savings

Customer satisfaction—Feedback from user departments, whether obtained informally or through more structured questionnaires and surveys, can be a good source of data on the functioning of a purchasing department and how well it is serving its customers.

Supplier Selection and Evaluation[3]

To effect the collaborative relationships necessary to make the supply team work effectively, the number of overall suppliers is drastically reduced through the creation of "commodity contracts," by grouping similar items, through supplier capacity planning that incorporates supplier production capabilities, and by coordinating information systems which link the buying company's forecasting and master production schedules.

Top purchasing organizations typically select suppliers using a "team buying" process that combines an organization's technical skills and quality leadership with those of the customer, the supplier, and the negotiation and contract creating proficiencies of marketing and purchasing professionals. These buying teams combine their talents based on careful member selection, education, and training, and they merge contractual and technical expertise with individuals certified in their specialties—each with a dedicated customer focus. Management establishes productivity targets and monitors supplier management through exception-based internal and external benchmarks and dollar-based reviews to ensure competitive supplier selections.

Negotiations with existing and potential suppliers are very deliberate and often time-consuming to ensure that mutually beneficial alliances are forged. Frequently, profits are "shared" using "product gross margin" negotiation procedures. Suppliers are frequently selected using a "Least Total Cost" computer-based technique that considers not only the lowest

---

[3]Many world class supplier selection and evaluation practices were adapted from empirical research conducted by Robert L. Janson, C.P.M., C.P.I.M, Senior Manager, Ernst & Young.

National Association of Purchasing Management®

Thank you for your order.

If you would like more information about our programs and services, please complete this card and drop it in the mail.

Name ____________________

Address ____________________

City ____________________ State ________ Zip __________

Yes, please send me information about:

- ☐ Membership
- ☐ Certification
- ☐ Seminars
- ☐ Educational Products

National Association of Purchasing Management®

Thank you for your order.

If you would like more information about our programs and services, please complete this card and drop it in the mail.

Name ____________________

Address ____________________

City ____________________ State ________ Zip __________

Yes, please send me information about:

- ☐ Membership
- ☐ Certification
- ☐ Seminars
- ☐ Educational Products

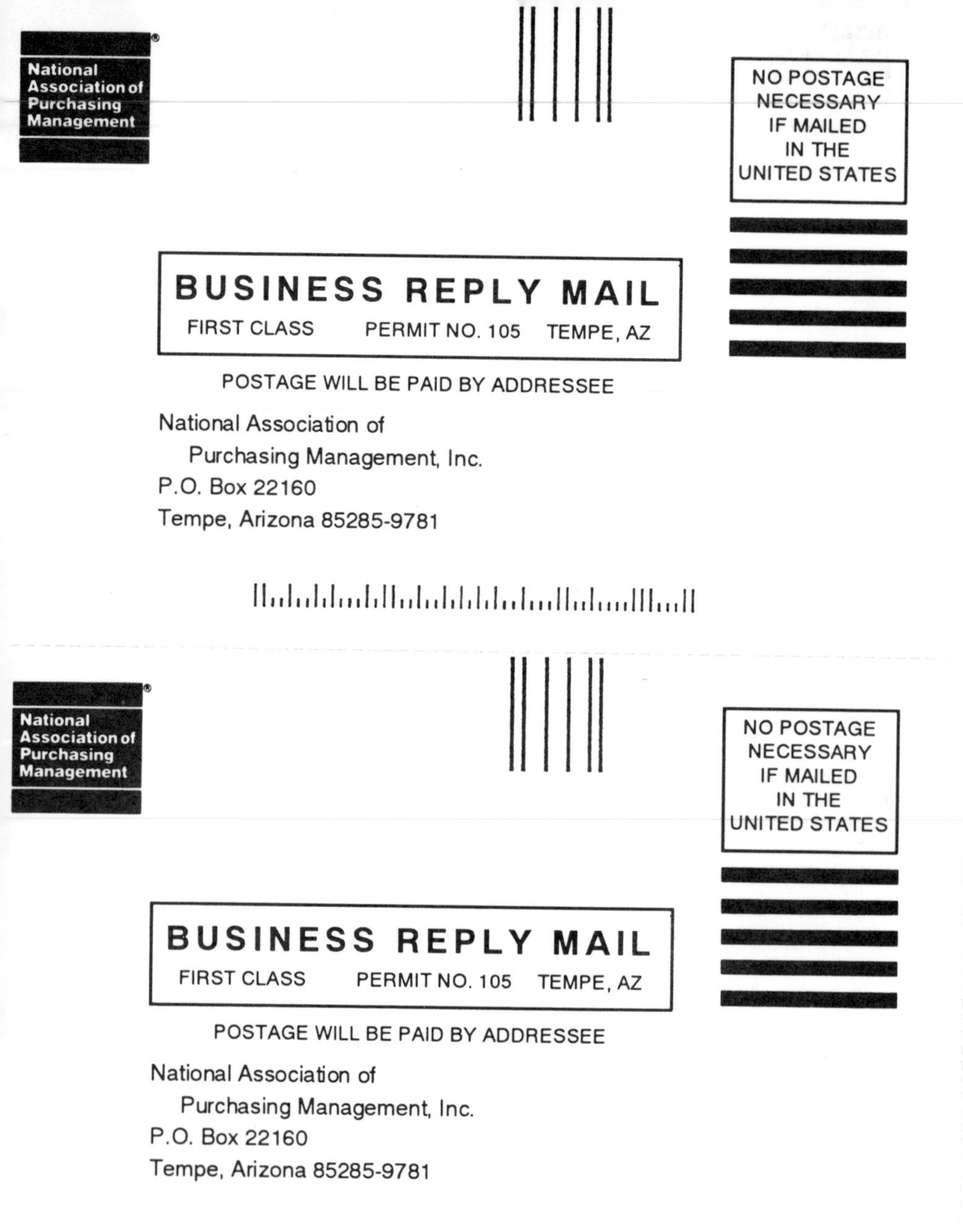

National Association of Purchasing Management®

NO POSTAGE NECESSARY IF MAILED IN THE UNITED STATES

BUSINESS REPLY MAIL

FIRST CLASS PERMIT NO. 105 TEMPE, AZ

POSTAGE WILL BE PAID BY ADDRESSEE

National Association of
Purchasing Management, Inc.
P.O. Box 22160
Tempe, Arizona 85285-9781

National Association of Purchasing Management®

NO POSTAGE NECESSARY IF MAILED IN THE UNITED STATES

BUSINESS REPLY MAIL

FIRST CLASS PERMIT NO. 105 TEMPE, AZ

POSTAGE WILL BE PAID BY ADDRESSEE

National Association of
Purchasing Management, Inc.
P.O. Box 22160
Tempe, Arizona 85285-9781

price, but also: (1) the ability to provide quality products that meet or exceed customer needs; (2) the capability to deliver frequently to point-of-use; or (3) the capacity to deliver added value through engineering design and supplier performance or responsiveness. Typically, purchase agreements are for three to five years, with incremental price adjustments keyed to governmental indices or other mutually acceptable formula-based measures that are reconciled annually.

Cost reductions are frequently achieved through continuous joint efforts. Higher volume suppliers typically assign full-time employees "on site" within the buying organization, thereby eliminating the need for frequent sales calls. Training sessions are held for both company and supplier personnel—both separately and jointly—to foster a spirit of partnering and also to communicate supply management processes and expectations.

Internal Audits/Self-Governance

A purchasing audit is a comprehensive, systematic, independent, and periodic examination of a company's purchasing environment, objectives, strategies, and activities, with the intention of identifying strengths and weaknesses, and developing a plan of action to improve purchasing performance. Ideally, audits are conducted by outside consultants to ensure the necessary objectivity and independence of judgment. They are, however, conducted internally as well. The drawback to this process may be a hesitancy on the part of the auditors to be critical of performance even when justified.

Exercising Management Control in Response to Results

Evaluation should be an ongoing process, a catalyst for improvement in the purchasing function, and a means of validating performance to management's expectations. Once the results of a performance appraisal are obtained, management must be sincere in its efforts to correct the problems identified through the appraisal. Otherwise, the entire performance appraisal process will serve no useful purpose.

Comparing Performance with the Plan

The individual and his or her supervisor jointly review the performance and compare it with the plan (or standard). Performance must be evaluated as objectively as possible in light of goals and expectations established in the prior planning process. From the results of this planning and evaluation process, subsequent plans can be developed to move toward further achievements in terms of both departmental and individual progress.

Reestablish Department Objectives

The final step after determining the relationship between activities and results is to reestablish the objectives based on the lessons learned from undertaking the department-level evaluation. The process is cyclical and requires a continual reevaluation of the main steps.

REPORTING MATERIALS MANAGEMENT RESULTS

Probably one of the most unpalatable tasks that materials managers and purchasing professionals have is making a written or oral report to top management about the functioning of their departments. In fact, many professionals find report writing to be tedious and avoid it if they can for as long as they can. Oral reports are often dreaded more so than written ones. This is especially true if the oral report is made to a group of administrators. A survey found the number one fear of the average North American was the fear of speaking in public. In fact, speaking in public was rated as the number one dread over such other choices as being audited by the IRS, dying, or going to the dentist. In any case, what to report, how often, and how to report are decisions of paramount importance to purchasing professionals. Effective writing and speaking is predicated on understanding the audience and succinctly conveying the intended message. A basic checklist for facilitating this process is provided in Table 10.4. These steps are not necessarily in sequence and should be tailored for each report.

What to Report

One aspect of report or speech writing that consumes time initially is deciding what to write about. In general, purchasing's operating reports which are prepared on a regular basis—monthly, quarterly, semiannually, or annually—often include: market and economic conditions and price performance; inventory investment changes; purchasing operations and effectiveness; and operations affecting administration and financial

TABLE 10.4
Communications Checklist

Communications Checklist
Analyze the purpose & audience
Conduct the research
Support the ideas
Get organized
Draft & edit using active language
Actively solicit feedback

activities.[4] Table 10.5, consisting of 27 common report items, is provided to help overcome the initial hesitancies associated with writing reports. Although not an exhaustive list, it serves as a starting point and contains many of the frequently required items.

One of the most effective tools in streamlining the supply management process, however, is to ask continually, "Is it necessary?" Often, the need to communicate no longer exists or the communication can occur less formally. Purchasing professionals must aggressively seek opportunities to reduce procedural, batched (and often vertical) written and oral communications, supplementing them with less formal and more frequent (typically horizontal) interactions at all levels of the organization and between members of the supply team. If the need to communicate remains, it is necessary to ascertain the purpose for the communication,

[4]Leenders and Fearon, p. 467.

TABLE 10.5
Commonly Reported Purchasing/Materials Management Items

	Monthly	Year To Date	Last Year To Date	Variance This Year To Date	Annual Summary
Cost Savings					
Savings as a result of make or buy decisions	X	X	X	X	X
Savings as a result of value analysis/standardization	X	X	X	X	X
Department Performance					
Number of purchase orders issued	X	X	X	X	X
Number of rush orders processed	X	X	X	X	X
Total purchases in dollars	X	X	X	X	X
Ratio of total dollars purchased to total revenue received	X	X	X	X	X
Average cost of writing a purchase order	X	X	X	X	X
Average cost of buying $1 of materials or services	X	X	X	X	X
Research results—progress and/or conclusion	X				X
Summary of years high points					X
Inventory Control					
Average inventory—total dollars	X	X	X	X	X
Average inventory per bed in dollars	X	X	X	X	X
Turnover rate by commodity group	X	X	X	X	X
Market Conditions					
Price trends	X	X	X	X	X
Lead time—average delivery time estimated/weeks	X				X
Shortages current and probable in future	X				X
Budget Performance					
Operating—actual expenses vs. budget	X	X	X	X	X
Short explanation of any significant variances	X	X			X
Capital expenditure, progress, and difficulties	X	X			X
Payments					
Ratio of orders with cash discount (2/10/net 30) to total	X	X	X	X	X
Ratio of cash discounts offered versus cash discounts taken	X	X	X	X	X
Receiving and Stores					
Percentage of accuracy of inventory on 10 randomly selected items	X	X	X	X	X
Average number of incoming shipments	X	X	X	X	X
Average number of requisitions processed	X	X	X	X	X
Vendor Performance					
Percent of on-time delivery	X	X	X	X	X
Percent of rejected items	X	X	X	X	X

which can usually be characterized as one of three purposes that often overlap: to direct, to inform (or to question), or to persuade. Once the need and purpose for the communication is established, it is a good idea to examine the list of suggested topics, eliminate the ones that are not applicable, and add any others that are needed. After the tailored list is completed, it is necessary to gather the requisite information.

The Written Report Outline

In writing a letter or report it is a good idea to build an outline as a first step. An outline helps ensure that all the important points are covered, serves as a foundation for the requisite research, and helps to organize the resulting report logically. A standard outline for report writing might include:

- ***Subject:*** A one-sentence statement that not only gives the reader an idea of what the report is about, but also tries to pique the reader's interest.
- ***Summary:*** Usually placed at the end of the report because it usually follows logically from the data presented before it. However, the summary is placed first because that is what most people want to know. The rest of the report is used to support the summary. The summary then should be a concise presentation of the data or "meat" of the report. For example:

  Summary: An estimated annual $7,735 savings in labor and transportation cost can be realized by changing the mode of handling and hauling items from outside storage:

A. Direct labor	$3,900
B. Transportation	2,990
C. Warehouse charge	845
D. Total annual savings	$7,735

- ***Recommendations:*** If there are any recommendations to be made, they should follow the summary. The recommendation should suggest a specific action. You may want to make one or more recommendations. Each recommendation should be a one sentence *specific* statement that tells what you think should be done. Do not attempt to defend or sell your recommendation in this section, because the next section is reserved for that purpose.
- ***Discussion:*** A discussion is a critical examination by argument. In other words, the purpose of this section is to present both *pro* (for) and

TABLE 10.6
Characteristics of Effective Reports

Characteristics of Effective Reports
Title clearly describing the nature of the report
Brief summary of key information
Conclusions (beginning of the report)
Recommendations (end of the report)
Short statistical tabulations, charts, or graphs (body of the report)
Lengthy statistical tabulations (in appendix with analysis in the body of the report)

con (against) arguments with regard to how the recommendations will or will not satisfy the issue involved. Obviously the pros should outweigh the cons in any successful defense of a recommendation.

• ***Calculations:*** Calculation means to compute or estimate by mathematics. The purpose of this section is to show the mathematical process used to derive estimated savings, expected profits or losses, or other pertinent data. If no figures are presented in the report that need support, this section should be eliminated.

As shown in Table 10.6, an effective report includes key attributes. The following ideas may also help in writing or delivering reports:

1. Keep it short—one or two typed pages.
2. Short sentences—average length about 17 words.
3. Use plain language—common words.
4. Use visual aids when applicable, such as pictures, graphs, charts, and tables.

TABLE 10.7
Six Signals All Audiences Want to Hear

Six Signals All Audiences Want to Hear
I will *not* waste your time
I know who *you* are
I am well *organized*
I *know* my subject
Here is my most *important* point
I am *finished*

Source: Ed Wohlmuth, *The Overnight Guide to Public Speaking,* Philadelphia: Running Press, 1983, p. 28.

Oral Reports

As mentioned previously, managing the fear of speaking is often the most difficult hurdle to overcome. One of the best ways to overcome this fear is to be well prepared. Most important, the speaker must prepare with the audience in mind. Six signals that all audiences want to hear are provided in Table 10.7. Additionally, if the only knowledge of the subject is in the speaker's notes, most likely he or she is insufficiently prepared. In preparing to deliver an oral report one should learn much more (five times more) about the subject than one plans to use. This technique not only adds confidence but also helps during question and answer periods.

The second method to gain confidence is practice. This, of course, requires a lot of time and effort. However, if oral reports are an important part of the job, then the effort will be worthwhile. Additionally, many purchasing professionals find it useful to take speech courses or to join Toastmasters International to get the experience and coaching needed; but in the final analysis, the best way to improve public speaking is to take advantage of every opportunity to get in front of a crowd.

Most of the suggestions above that relate to written reports apply equally to oral reports. Additional tips include:

1. Start with an outline
2. Keep it short
3. Use visual aids (overheads, slides, films, TV tapes, charts, or handouts)
4. Be very organized so that things flow smoothly
5. Use anecdotes (a short account of an interesting or humorous incident) that either explain a point or otherwise help to make a point

Remember, however, that the most important facets of report generation are (1) reporting the right information and (2) ensuring the accuracy and validity of the information—so it often is wise to check and recheck the facts.

KEY POINTS

1. Control is the process of establishing standards of performance, evaluating actual performance against these standards, and an indicating when corrective action needs to be taken. Control softens the impact of unforseeable problems on purchasing performance.
2. Effective purchasing controls are enacted in conjunction with purchasing planning and contribute significantly to making the supply team work. A purchasing manager should develop systems to control any plans or programs for which he or she is accountable. The effects of neglecting control for issues such as operating costs can be dramatic and disastrous.
3. When developing a system to control a task, the manager must plan or determine how to control the process. All the tools required must be organized ahead of time. Employees should be directed to implement the controls, then be empowered and motivated to monitor the controls without absolute managerial supervision.

4. Evaluation is an important consideration for purchasing managers in monitoring progress. Essential to adding value is the continual evaluation of the organization and its supply team partners, purchasing and its organizational counterparts and individuals, with the intent of continuous improvement.
5. When control techniques reveal that a plan has failed or has a flaw, the manager should replan and reorganize to correct the situation. This process can be repeated many times in the process of fine-tuning a plan.
6. When performance deviates from plan, the timing and nature of the corrective action depends on the magnitude of the problem. Sources for corrective action include poor or unrealistic planning, poor organization, poor direction and motivation, and unforeseen changes in the plan's operating environment.
7. Purchasing department appraisals and controls ensure that the department is conforming to and supporting the organization's mission. A manager must rate purchasing's effectiveness and make improvements where necessary.
8. When developing a system to control and evaluate a department, a manager should analyze the department's objectives and performance targets.
9. Factors for evaluation must be derived and ranked in order of importance. Factors such as prices paid for materials, material quality, and supplier reliability can all ensure that maximum value is derived from supplier relationships.
10. Purchasing audits provide an unbiased examination of the department's objectives, strategies, results, strengths, and weaknesses. Managers must act on the auditor's recommendations to correct the problems identified.
11. Many issues must be considered before entering into a strategic relationship, and purchasing professionals have to exhibit the necessary skills to effectively evaluate and control supply team partners.
12. Appraisal and evaluation eventually lead to management reports. Deciding what to report, when, and how to report are key decisions managers must make.
13. When deciding what to include in a report, a manager must decide the level of detail at which to report and the topics that

need to be covered. A manager can focus the report by asking "Is it necessary?" when considering topics and details to be included.

14. The written report should begin as an outline of the important points that need to be covered. A typical report consists of a subject, a summary of the relevant facts, recommendations, a discussion of the alternatives, and an explanation of how estimates and projections were calculated.
15. Although public speaking ranks high on most people's anxiety list, the fear of giving oral reports can be reduced by being well prepared and knowledgeable about the topic. Keep the report short, organized, and accurate.
16. Effective written and oral communications that serve a useful purpose also play a significant role in controlling and evaluating. To add value within the supply chain, purchasing professionals must be able to communicate succinctly.
17. Frequent, horizontal, informal communications are important to streamlining supply management. Clarity, conciseness, and accuracy are fundamental hallmarks of effective reports.

REFERENCES AND RECOMMENDED READINGS

Badawy, M.K., *Developing Managerial Skills in Engineers and Scientists: Succeeding as a Technical Manager,* New York: Van Nostrand, 1982.

CHAPTER 11

THE ROLE OF THE PURCHASING PROFESSIONAL: THE CHALLENGES AHEAD

INTRODUCTION

Throughout this book we suggest that purchasing professionals need to become more strategic in their views and more proactive in their contributions to the corporate planning process. As discussed earlier, purchasing must exercise a significant role in developing a firm's competitive advantage, particularly with regard to development lead time, quality, and total systems cost. The strategic firm must include purchasing and supply considerations in its planning, development, and operations.[1] While such efforts make prudent business sense, other challenges are facing senior management. North American business cannot enter the next century with outmoded structures and processes.

This chapter has been adapted from work conducted in conjunction with Robert E. Spekman, Darden Graduate School of Business, University of Virginia and Deborah J. Salmond, University of Baltimore. A more comprehensive version entitled "At Last Purchasing is Becoming Strategic" written by Robert E. Spekman, John W. Kamauff and Deborah J. Salmond is available in Long Range Planning, Vol. 27, No. 2, April, 1994, pp. 76-84.

[1] For additional information on the strategic role of purchasing, please see Ellram, Lisa M., "The Supplier Selection Decision in Strategic Partnerships," Journal of Purchasing and Materials Management, Fall 1990, pp. 8-14; Freeman, Virginia T. and Joseph L. Cavinato, "Fitting Purchasing to the Strategic Firm: Frameworks, Processes, and Values," Journal of Purchasing and Materials Management, Winter 1990, pp. 6-10; Landeros, Robert and Robert M. Monczka, "Cooperative Buyer/Seller Relationships and a Firm's Competitive Posture," Journal of Purchasing and Materials Management, Fall 1989, p. 10. Most recently, Monczka's position has been presented in Landeros, Robert, and Robert Monczka (1990) and Spekman, Robert, (1981) and Spekman, Robert E., "Strategic Supplier Selection: Understanding Long-Term Buyer Relationships," Business Horizons, Vol. 31, Number 4 (July-August 1988).

The practice of management is embarking on a new plateau and is developing a series of concepts and techniques for further improving corporate performance into the 21st century. One cannot open a business periodical without reading about mass customization, business process redesign, lean production, the learning organization, or corporate reengineering as radical new visions for changing the direction of business practice. An example of these changes affecting purchasing is in available plant capacity. According to Robert L. Janson, senior manager, Ernst & Young, "On average, the typical manufacturer now has plus or minus seven percent flexibility in capacity. Five years ago, the inherent flexibility was 15 to 16 percent." At each turn, organizations are challenged by the threat of "the new competition,"[2] and survival under such constraints will require different perspectives and skills.

These fundamental and profound changes in business practice have significant implications for the purchasing profession. This chapter examines these changes as they apply to making the supply team work and offers insights into how the supply management activities of the firm are likely to be affected. In addition, this chapter continues the array of suitable responses for purchasing professionals so they can continue to be strategic assets for their firms. We begin with a discussion of both the components of the new competition and the tenets of business reengineering and mass customization, followed by a look at the implications for the purchasing professional. Finally, enablers and inhibitors of change are described.

The New Competition

The term "new competition" has been discussed in other recent works on North American economic development since World War II,[3] but Michael Best states that the new competition consists of four dimensions: organization of the firm; types of coordination across phases of the production chain; organization of the sector; and patterns of industrial policy. The basic premise of the new competition is that firms will no longer compete in ways we have witnessed in the past. The implications of the new competition on purchasing are shown in Table 11.1.

In this context, bureaucratic and hierarchical firms are doomed. The new competition embodies global networks, at the core of which are adaptive,

---

[2] Best, Michael, *The New Competition,* Harvard University Press, Cambridge, MA: 1990.

[3] Piore, Michael and Charles Sabel, *The Second Industrial Divide,* New York: Basic Books, 1984.

TABLE 11.1
The Impact of the "New Competition" on Purchasing

DIMENSIONS	PURCHASING IMPACT
Organization of the Firm	Global networks whose core is based on flexible, creative learning organizations Designed to meet the needs of the marketplace
Types of Coordination	Breakdown of functional silos Viewing functional units as interdependent parts Coordinating the supply value chain for competitive advantage
Sector Organization	Removal of the complex tension between firms (strategic relationships versus competing within the supply chain)
Patterns of Industrial Policy	Structural transformations beyond natural functional barriers evolving into complex inter-firm relationships Redefining relationships with suppliers, customers & even competitors

learning organizations that are flexible enough to respond quickly to marketplace changes. Micro-level concerns are of particular interest to purchasing.

Types of coordination refers to breaking down functional groupings and viewing functional units as interdependent parts of the firm. This coordination also incorporates the relationship between buyers and suppliers and the need to coordinate the supply chain. What evolves is a network in which firms are linked through a series of connections whose primary objective is to gain sustainable strategic advantage.

Sector organization refers to the complex tension between firms, since they can simultaneously interact as allies and competitors. These two dimensions imply a strategy whereby firms acknowledge at the outset that it is often impossible to "go it alone" (e.g., the mutually beneficial relationship between Hewlett-Packard and Intel). In addition, the structural transformation required to face the "new competition" extends beyond the natural functional boundaries (and barriers) of the firm and naturally subsumes a complex web of inter-firm relationships that span the firm's strategic activities.

In the spirit of the new competition, firms redefine their relationships with suppliers, customers, and even competitors, as well as restructure their internal organization. Many authors have attempted to explode the myth of the cooperative single source, to the extent of suggesting that many apparently cooperative, egalitarian relationships in Japan are actually deals made between unequals, with the buyers wielding huge amounts of power over their single sources.[4] Although these authors and others are quick to refer to the risks associated with closer relationships between buyers and sellers, there exists a large and compelling body of evidence suggesting that in response to the global challenges facing North American manufacturers, cooperative relationships between buyers and suppliers offer a potential source of competitive advantage.[5] The debate here should not center on whether strategic alliances are a good idea; but rather on how

---

[4]For additional information on single sourcing, please see Ammer, Dean S., *Materials Management and Purchasing,* Fourth Edition, Homewood, Illinois: Irwin, 1980, and Ishikawa, Kaoru, *What is Total Quality: The Japanese Way,* translated by David J. Wu, Englewood Cliffs: Prentice-Hall, Inc., 1985; Leenders and Blenkhorn, p. 140; McMillan, John, *"Managing Suppliers: Incentive Systems in Japanese and U.S. Industry," California Management Review,* Summer 1990, pp. 38-55; Newman, Richard G., *"Single Source Qualification," Journal of Purchasing and Materials Management,* Summer 1988, pp.10-17, and Newman, Richard G., *"Single Sourcing: Short-Term Savings Versus Long-Term Problems," Journal of Purchasing and Materials Management,* Summer 1989, pp. 20-25..

[5]Bartholomew, Dean, *"The Vendor-Customer Relationship Today," Production and Inventory Management Journal,* Volume 25, Number 2, 1984, pp. 106-121; Jackson, B., *"Build Customer Relationships that Last," Harvard Business Review,* 6, November-December 1985, pp. 120-128; Bhote, Keki R., *Supply Management: How to Make U.S. Suppliers Competitive.* New York: American Management Association Membership Publications Division, 1987, John, George and B. Weitz, *"Forward Integration into Distribution: Empirical Testing of Transaction Cost Analysis," Journal of Law, Economics and Organization,* 4 (2), 1988; Heide, Jan B. and George John, *"Alliances in Industrial Purchasing: The Determinants of Joint Action in Buyer-Supplier Relationships," Journal of Marketing Research,* Vol. XXVII, February 1990, p. 25; and Spekman, Robert E. and Kirti Sawhney, *"Toward a Conceptual Understanding of the Antecedents of Strategic Alliances,"* in Wilson, D. T. and K.E. Christian Möller (eds.), *Business Marketing: An Interaction and Network Perspective,* Boston, MA: Kent Publishing, 1990.

best to manage and leverage the skills and talents of the ally. The purchasing manager, who must serve as a broker of inter-firm information exchanges, as opposed to being a transaction accountant, becomes a critical participant in the process, guiding both the formation and implementation of strategic relationships and inter-firm supply networks. Thus, the purchasing professional becomes an information manager and manager of external manufacturing, with responsibility throughout the supply chain; he or she must gather and filter relevant procurement-related knowledge about products, processes, competition, and other factors that can affect the firm's competitive posture. The shift harkens a transformation from reactive, transaction-based purchasing to proactive, value-added supply chain management.

Business Reengineering[6]

Business reengineering refers to a process, a view of the firm in which all the old rules are out and managers are charged with finding new ways to manage the business. It is important to note the difference between business reengineering and benchmarking. Benchmarking, as practiced by most firms, is merely the pursuit of benchmarks; it often is predicated on shortcuts and it results in attempts to find the company that performs a particular task or function best and ultimately to emulate that firm.[7] Strategic benchmarking emphasizing a mutually beneficial exchange of information between partners[8] takes the process of business reengineering one step further by addressing the fundamental nature of the task or function and questioning whether it is worthwhile to even engage in such an activity. Reengineering acknowledges that being "world class" at a particular activity is laudable and probably a worthy pursuit; however, one should not seek such a goal blindly. One should question the basic processes underlying the task or business activity. It would appear that both concepts are complementary, and that benchmarking results can provide useful input to a business process redesign activity and to the evolution of the so-called learning organization.

---

[6]The majority of this discussion is based on Hammer, Michael and James Champy, *Reengineering the Corporation: A Manifesto for Business Revolution,* New York: Harper Business, 1993.

[7]Kramer, Steven B. and John W. Kamauff, Jr. *"Benchmarking: An Analysis of Benchmarking for Implementation at Westinghouse Electronic Systems," Darden School Working Paper (DSWP-93-09),* 1993.

[8]Kramer, Steven B. and John W. Kamauff, Jr. *"Facets of Benchmarking," Darden School Working Paper (DSWP-93-08),* 1993.

According to Percy Barnovik, CEO of Asea Brown Boveri (ABB), "benchmarking is essential for global competitiveness." However, by its very nature as a window on the external world, benchmarking affords an excellent opportunity for purchasing professionals to play an important role in contributing to this process. By exercising the inherent access to information from outside the firm, purchasing professionals can effect meaningful benchmarking alliances.

Similar to the notion of functional coordination raised above, business reengineering does not take the singular functional unit as sacrosanct. At the core of the process is a recognition that functions must cooperate and share in the responsibility of meeting the needs of the customer. A firm is unlikely to have a "purchasing problem" or a "production problem." In this regard, concurrent engineering or integrated product development have become widely accepted means for facilitating work flow between design and engineering.[9] Similarly, it is more likely that most supply problems transcend the purchasing function. For example, in a situation in which costs are out of line, the goal would be to have all relevant parties, internal and external, work together to solve the problem—in essence, concurrent procurement. This process would advocate that target costs should not be dictated to suppliers, as they would not be to internal departments. Instead, to make the supply team work, the firm should help the supplier to achieve targets over time and should be willing to support the supplier with engineering support, technical assistance, or process innovation. Upon investigation it might be found that earlier supplier involvement in the design stage can reduce the cost of the component, or that greater information sharing permits the supplier to better manage its logistics and production scheduling, thereby also reducing costs. In fact, Boeing Commercial Airplanes has adopted a design-build team approach that includes both key suppliers and customers throughout the project life cycle for its 777 program. With clear advantages in terms of time and innovation, such changes have and will continue to have a profound impact on the purchasing profession.

---

[9]Clark, Kim B. and Takahiro Fujimoto, *Product Development Performance: Strategy, Organization, and Management in the World Auto Industry,* Boston: Harvard Business School Press, 1991; Rosenthal, Stephen R., *Effective Product Design and Development: How to Cut Lead Time and Increase Customer Satisfaction,* Homewood, IL: Business One Irwin, 1992 and Wheelwright, Steven C., and Kim B. Clark, *Revolutionizing Product Development: Quantum Leaps in Speed, Efficiency and Quality,* New York: Free Press, 1992.

Mass Customization

It is difficult to discuss the new competition or reengineering without addressing the notion of mass customization, since the three concepts are intricately linked and all have a profound effect on the procurement activities of the firm. The concept of mass customization[10] evolves from a failure inherent in the mass production paradigm developed and espoused by bureaucratic organizations. While low costs are important and are a fundamental goal of the mass production paradigm, across a number of industries the premise upon which mass production is built is no longer valid. Markets are not homogeneous—they tend to be fragmented; product life cycles are shorter and cycle time has become a critical element of competitive response[11]; and customers are demanding tailored products and services. Where mass production provided efficiency through stability and control, mass customization seeks variety and customization (at low costs) through flexibility and quick response. This new paradigm demands processes, systems, and structures that are responsive to changes in the marketplace and that can be implemented with little change in the cost structure of the product and/or service.

Mass customization attempts to achieve the synthesis of two competing management forms. Low cost is achieved through both economies of scale and economies of scope. The goal is to minimize the penalty of individuality, since unit costs fall with the increased volume achieved through the entire system. To be sure, there are limits to the degree of variety but the range is such that, depending on the product, a vast number of options are often available. For example, Motorola can robotically assemble thousands of different pagers during a single shift with zero set-up cost and with a true lot size of one.[12]

It should come as no surprise that mass customization is a result of the new competition and is a function of the firm's ability to link product and process development. In fact, the process of business reengineering lies at the core of successful mass customization. That is, a focus on fundamental processes allows a better grasp of the basic changes that affect

---

[10]Much of this discussion is based on work by Pine, Joseph B., *Mass Customization,* Cambridge, MA: Harvard Business School Press, 1993.

[11]Stalk, George, Jr. and Thomas M. Hout, *Competing Against Time: How Time-Based Competition is Reshaping Global Markets,* New York: The Free Press, 1990.

[12]Pine, p. 146.

the firm's ability to compete and to meet the ever-changing needs in the marketplace. According to John Marous, former chairman of the Westinghouse Electric Corporation:

> Processes are at least as important as products because it's in the process where many of the biggest competitive gains can be made; gains in world-class quality; gains in cost; gains in global market flexibility. At Westinghouse, we believe we understand this shift in perspective. In fact, it's something we must do, given today's global pressures and our dynamic, rapidly changing environment.[13]

For the purchasing manager, it is critical to understand the whole of the value chain[14] and to appreciate the full range of activities that comprise the manufacturing of a component or finished product. It is again clear that functional integration takes precedence over individual functions and that cross-functional teams rise in importance. The question facing the purchasing manager is one of how to assemble the internal resources to develop requirements for the best set of suppliers.

Organizational Transformation and Its Implications

To adapt to the new competition, fundamental transformations will occur at all levels of the firm and will affect each functional area. For purchasing, these profound changes are critical for the future well-being of the firm. A key enabler of the new competition is the organization's ability to form linkages with other businesses that can provide competitive advantage. The purchasing function can emerge as a key player in the process and can serve as the point around which both internal and external relationships are better nurtured and managed. Purchasing's revolutionary (as opposed to evolutionary) role in effecting this transformation is shown in Table 11.2.

Purchasing's Role in Knowledge Creation

To begin, senior management must engender a strong commitment to the firm's mission and goals. Management must not only articulate what the goals are but must also have a well-defined plan for accomplishing these

---

[13]Marous, John quoted in "Know Your Process (And Manage It)," Revision 18, Westinghouse Electric Corporation, 1990, p. 6.

[14]For a full discussion of value chain analysis, see Michael Porter, *Competitive Advantage: Creating and Sustaining Superior Performance,* New York: The Free Press, 1985.

TABLE 11.2
Purchasing's Revolutionary Role in "New Competition"

EVOLVING ROLL	REVOLUTIONARY ROLL
Transaction accountant	Information exchange broker
Administers inter-firm contracts	Guides the formation & implementation of strategic relationships & inter-firm supply networks
Primary POC with suppliers	Manager of external manufacturing
Interface with first-tier suppliers	Responsibilities throughout the supplier value chain
Minimizes risk (e.g., disruption of supply, incoming defects, etc.) to the buying organization	Manages and leverages the skills of the supply chain
Competitor analyses	Strategic benchmarking of the supply chain
Continuously improving the existing procurement process	Questioning the supply chain process (reengineering procurement)
Reacting to external stimuli (reactionary change)	Proactively accessing external information (proactive change)
Concurrent engineering	Concurrent procurement
Safeguarding proprietary/critical information (transaction driven)	Enhanced information sharing throughout the value chain (early supplier involvement)
Unidirectional communication	Simultaneous two-way communication
Cross-functional coordination	Functional integration
Cause-and-effect problem solving	Systems thinking
Purchasing mentality	World view

goals. Execution is essential. Each functional unit must understand precisely what its role is in the process and sublimate its agenda for the greater purpose—sustained competitive advantage. For the purchasing manager, it requires a vision that spans the entire value chain, beginning with purchased materials and ending with the customer. This vision must be congruent with and in support of the firm's mission and goals.[15] Stuart suggests that such consensus is no easy task and that achieving corporate goals within a framework of personal satisfaction remains a continuing management challenge.[16]

A purchasing organization that is driven by outmoded measures of efficiency, such as purchasing orders per buyer, and is rewarded based on a singular focus on cost, is out of touch with the new competition and is, over the long-term, a detriment to its firm. Such a focus fails to acknowledge that many of these expense-incurring activities are being automated, and information technology is replacing these more mundane considerations. In a recent benchmarking effort designed to facilitate their gravitation toward commercial sourcing practices, Westinghouse Electronic Systems Group, a long-standing and highly regarded defense contractor, learned quickly that its purchasing metrics were outmoded. By reexamining its underlying purchasing processes and comparing them with leading-edge practitioners, Westinghouse was able to implement quickly a more viable commercial procurement system. EDI, and other systems intended to link firms electronically, have begun to change fundamentally many purchasing activities. Workflow monitoring, inventory control, and routine order processing are now computer driven and require very little managerial attention.

Instead, purchasing must devote greater attention to those skills inherent in understanding and contributing to the firm's value-creating activities. In fact, sourcing contributes to cost control as well as to improving overall performance in quality, dependability, flexibility, and innovation.[17] Strategic sourcing is quite simply this overall process of creating a value-adding or optimal mix of supply relationships to provide a competitive advantage. In this context, the purchasing manager becomes a source of knowledge regarding the processes that comprise the value chain. Working

---

[15]Hayes, Robert H. and Steven C. Wheelwright, Restoring *Our Competitive Edge: Competing through Manufacturing,* New York: Wiley & Sons, 1984.

[16]Stuart, F. Ian, *"Purchasing in an R&D Environment," Journal of Purchasing and Materials Management,* Fall 1991, pp. 29-34.

[17]Hayes, Robert H., Steven C. Wheelwright and Kim B. Clark, *Dynamic Manufacturing: Creating the Learning Organization,* New York: The Free Press, 1988.

in conjunction with other functional managers and as part of a learning organization, purchasing provides input to make-or-buy or strategic relationship decisions. Whether the corporate objective is to pursue a strategy of cost leadership or differentiation, the role purchasing can play influences the firm's strategic posture.

Questions transcend the more parochial issue of whether a component or sub-assembly should be made or bought. Purchasing should actively contribute to the store of knowledge from which the firm defines its core competencies and decides which set of skills and activities to keep and which ones to shed. This "disintegration" of the firm typifies a key component of the new competition. Nonessential activities will flow from the firm to its network and this resultant web of firms will behave as though the work were done internally. Given purchasing's understanding of the supply market, it can help assess the leverage points that contribute to competitive advantage.

Making the supply team work does not, however, suggest "giving away the farm." Purchasing also plays a role in protecting the firm's competitive advantage, particularly as it relates to the body of corporate knowledge affected by the strategic relationship. Purchasing must help establish the parameters of information sharing and knowledge transfer between the partners. One must take care not to jeopardize the firm's core skills and capabilities by outsourcing key process or product elements and thereby losing that skill.[18] One must also be watchful of partners who learn at your expense and gain information that places the firm at a competitive disadvantage. For example, a supplier could learn information that permits it to compete with you at a point further downstream.

Purchasing as Account Manager

Throughout this book, we allude to the importance of partnerships and imply a transformation in how purchasing should approach its suppliers. We also express a vision of a network of companies in which information passes freely and ubiquitously. Clearly, purchasing plays a principal role in managing the array of suppliers who comprise that network. Earlier work has developed the notion of supplier selection and the criteria by which a reliable set of partners emerge from a potential set of suppliers, but the principal issue here is directing the key partnerships through their development and growth, *and* taking a proactive role in seeking, nurturing, and

---

[18]Davis, Edward W., "Global Outsourcing: Have U.S. Managers Thrown the Baby Out with the Bath Water?" Business Horizons, Volume 35, Number 4, July-August 1992, pp. 58-65.

managing these relationships. The level of analysis has shifted from the commodity class to the partner, or array of partners, who supply value to the firm's products/services as it attempts to meet the needs of its customers and its customers' customers. Purchasing not only manages the partner; it manages the range of possible support services the firm brings to the partnership as both attempt to work toward a common goal and set of objectives.

As the account manager, purchasing should be the primary contact between the firm and the supply partner through whom the initial scope and nature of the partnership is negotiated and implemented. Although the partnership might take on a life of its own in that information ebbs and flows across different functions and organizational levels, purchasing has a role in orchestrating the initial contact. Through the life of the relationship, purchasing helps develop and monitor the performance criteria upon which partnership effectiveness is based. In this gatekeeper role, purchasing is a key conduit through which certain partnership-relevant data ought to flow and be manipulated. Purchasing is a logical storehouse for partnership information with which the firm can better assess potential relationships and appraise ongoing relationships.

Purchasing's Role in Partnership Negotiations

Typically, supplier negotiations converge on such issues as price, quality, or delivery. In the new competition, negotiations are much more broadly defined and cover a much wider range of issues, such as cost reduction strategies over time, technology sharing and joint development, and issues germane to control and performance measures. The breadth and depth of these interactions require both strategic insight and business knowledge that typically exceed the range of information available to a purchasing manager. Because the results of these negotiations have immediate strategic importance for the firm, purchasing's responsibility extends beyond the range of issues usually found in more traditional market-based transactions. Additional/different interactive skills are also required for these expanded negotiations.

Beyond the substantive aspects of the negotiations, one must recognize that each member of the negotiation team shapes the tone for the partnership as it develops over time. Thus, the early phases of the negotiations in which partners' motivations are understood, expectations are shaped, and future interactions are outlined become critical to the successful outcome of the partnership. As is the case in all ongoing relationships, past behavior affects future interaction. The process of negotiations becomes very

TABLE 11.3
Transforming Purchasing

Transforming Purchasing
Engender a strong commitment to the firm's mission & goal
Help shift the firm's orientation to the customer & the marketplace
Broaden the purchasing vision to include the entire supplier value chain
Re examine the firm's underlying procurement processes and compare them with leading-edge practitioners
Devote greater attention to those skills inherent in understanding and contributing to the firm's value-creating activities
Work to influence the firm's strategic posture by actively contributing to the store of knowledge from which the firm defines its core competencies
Help establish the parameters of information sharing & knowledge transfer between partners
Exercise purchasing's information-brokering responsibilities to develop a complementary & proactive supplier network for the sole purpose of creating a sustainable competitive advantage

important. Notions of perspective taking, rapport building, trust, commitment, and information sharing become essential ingredients of partnership formation and nurturing. Purchasing managers must become much better skilled in these process-related areas.

Conclusion

Throughout this book, we have attempted to describe the new competition and present the implications it holds for the purchasing manager. To a large extent, this chapter is intended as a wake-up call. As businesses grapple with the way to make fundamental changes in how they compete, purchasing

TABLE 11.4
Transforming the Purchasing Professional

Transforming the Purchasing Professional
Manage both the partnerships themselves & the range of possible support services
Orchestrate the initial scope & nature of the partnership
Encourage the exchange of information within & across organizational boundaries at all levels of the respective firms
Help to develop & monitor the performance criteria upon which partnership effectiveness is based
Develop the requisite skills to nurture & sustain mutually beneficial relationships
Develop strategic insight & business knowledge transcending traditional purchasing information
Take calculated risks & explore alternatives in order to lead change in the organization
Create a mentoring culture within the firm

faces new challenges. Purchasing professionals must adapt to these changes and re-think the manner in which they formulate and execute purchasing strategy as a component of the larger corporate goals and objectives. Traditional views of suppliers and of other functional units are no longer valid. As functional and organizational boundaries fall, purchasing must acquire new skill sets and levels of expertise. But the responsibility for transforming cannot come from external sources; it remains resident within purchasing. To lead this transformation, purchasing professionals must change themselves.

The new competition raises the ante for success, and purchasing now faces new opportunities. Although we have seen marked improvement since the 1970s in both the level of professionalism and ability to think strategically, purchasing can no longer improve at a snail's pace. The dynamics of competition are such that speed is essential. Lead time, time-to-market, and cycle time are key metrics for the successful global player.

Notwithstanding these profound changes affecting the purchasing profession, the management cycle of planning, organizing, directing, motivating, controlling, and evaluating will remain fundamental to leveraging the value of purchasing in the organization. Undoubtedly, the inherent demands on people, processes, technology, and systems will be greater and the management cycle will necessarily have to be reiterated more frequently, so it is imperative for purchasing professionals to be proficient at these underlying management processes to truly make the supply team work effectively. The type of people we employ, the way we plan, organize, direct, motivate, control, and evaluate purchasing will have to reflect the new challenges and opportunities described herein. The challenges will be demanding but the rewards to ourselves and our organizations will clearly justify the sacrifices to be made.

KEY POINTS

1. North American companies are facing a period of dramatic changes in business practices. A purchasing department must adopt and accept these new techniques to remain strategic assets for their firm.
2. In the "new competition," companies can no longer compete using outdated techniques. Bureaucracy and hierarchy will make way for leaner, flexible, more creative management styles. Cooperation will replace conflict within the supply chain.
3. Business reengineering is the process of finding new ways to manage a business. Whereas benchmarking is the study and emulation of an ideal firm within an industry, business reengineering looks at the operation of tasks themselves and determines if and how a firm can become "world class" in the execution of those tasks.
4. The new competition has led to heterogeneous, fragmented markets. This has caused product life cycles to shorten and customization and flexibility in manufacturing to increase. The

purchasing manager must be able to develop the supply chain to comply with these new needs.

5. In the developing cooperative environment between companies, purchasing will become an important factor, as purchasing is usually the first and most common point of communication between companies.
6. Purchasing must develop the skills needed to contribute to the firm's value-creating activities. Instead of relying on old methods and benchmarking metrics, purchasing must become a knowledge base for its functions and skills in relation to the company's goals and competencies.
7. With the development of inter-company cooperation and supply chain partnerships, purchasing's role is to help nurture and manage these relationships. Purchasing should provide support to the partnership, while monitoring and controlling the flow of information between the companies.
8. With the shift in focus of negotiations from tactical to strategic issues, the role of the purchasing department has broadened to include partnership development as well as strategic policy development for the company.

INDEX

An n following a page number (as in 115n) refers to a footnote at the bottom of that page.